Praise for *Salts*

"*Salts* by La Toya Tanaka introduces salt as matter that begins in an electron transfer and grows into a force that shapes physiology, settlement, authority, belief, and technology. Tanaka shows how a simple bond dissolves into charged particles that move through organs and guide essential transport. She follows this motion into early extraction sites where steady access determines survival and directs the growth of towns. She then examines periods in which rulers use the control of salt and trade corridors to secure revenue and extend influence. Her study turns to healing practices in which measured mixtures support care and older traditions treat mineral balance as a marker of internal order. Tanaka concludes with modern research in which heated salts hold energy, steady experimental reactors, and support new storage methods that alter industrial planning. *Salts* by La Toya Tanaka earns respect through the force of its documented reasoning, demonstrated through cases that show how concrete decisions produced lasting outcomes. Her examination of the Kanawha producers is especially strong, since she uses their deliberate market flooding to show how coordinated pricing reshaped local production and pushed small operations out, giving the reader a clear line from action to consequence. Tanaka applies the same standard to political authority by tracing how the British Salt Act shaped daily conditions and set the stage for the 1930 march, presenting the event through records that show economic impact rather than relying on spectacle. The book's power comes from this consistent use of direct evidence, which gives each chapter a firm internal logic and shows how control of a basic mineral shaped public life across varied eras."

—Asher Syed, *Readers' Favorite*
5-Star Review

Salts

La Toya Tanaka

NEW YORK
Harvard Square Editions
2026

Salts

ISBN: 978-1-947386-03-7

1. Salts—Analysis and Crystallography—United States—Non-Fiction.
2. Salts—Mineral salts in the body and Nutrition—Non-Fiction.
3. Salt industry & trade—History—Non-Fiction.
4. Mines and mineral resources—Geology—United States—Non-Fiction.
5. Mineralogy—Salts—Analysis.
6. Chemistry, Inorganic—Salts—Textbooks.

Printed in the United States of America
Published in the United States by
Harvard Square Editions
www.harvardsquareeditions.org

Contents

I

The Secret of Salts

"Whoever possesses the salt that can be melted, and the oil that cannot be burned, may praise God."
—Arnaldus de Villanova
(c. 1235–1313 CE), Alchemist

Salts hold a secret that echoes through millennia. These simple crystals, so often taken for granted, turn out to be a microcosm of one of the oldest and most elegant concepts in the universe. From the ancient Chinese forces of yin and yang to the foundational principles of modern science, there has been a pervasive sense of two opposing forces seeking each other out for completion:

Viewing our planet's formation through the lens of an inhospitable gas giant like Jupiter, the appearance of salt must have been an admirable illustration of how volatile extremes can be balanced to bring stability on a planetary scale. Imagine nascent Earth's turbulent environment hosting two wildly unstable chemical forces: the highly reactive metal, sodium (Na), and a surging cloud of poisonous, greenish-yellow gas, chlorine (Cl). Each substance, in its pure state, represents an extreme form of instability. The sodium is a force of excess, defined by a valence electron it desperately needs to donate; the chlorine is a lackey—a force of deficit—defined by the lack of an electron it violently seeks to claim.

But when these two volatile forces meet within the planet's maelstrom, it's not bombardment, but tango-time. They engage in a fundamental act of mutual stabilization. The sodium sheds its electron in excess, and the chlorine receives the electron it lacks. This transfer neutralizes the extreme, inhospitable nature of both components, fusing them into a benign, crystalline compound: I-am-you-and-you-are-me salt. This is the fundamental law of nature in action. Two extremes resolving their internal tensions by bonding into a single molecular equalizer, salt ($NaCl$). A stable entity rooting a fundamentally unstable world. Rendering it habitable.

Seen through the lens of chemistry, salt is defined as the marriage between an acid and a base. Nature, in its pursuit of perfection, designed acids and bases to be a natural pair. According to another theory entitled "the Brønsted-Lowry Model," bases desperately seek a proton (H^+) to fill a void. Acids, on the other hand, are burdened with an extra proton (H^+) they are eager to shed. And this is how these complementary elements collide in a heavenly marriage to become a 'salt'.

This seemingly simple union of elements belies a profound impact when applied to medicine, for example. Anyone who's ever been admitted into a hospital knows that one of the first things a nurse will do is connect you to an intravenous (IV) line with a bag of salt (NaCl) fluid for rehydration. It is well acknowledged in the pharmaceutical industry that about half of all drugs contain salts, a fact re-emphasized in *European Pharmaceutical Review's* "Pharmaceutical salts of small molecule drugs." Not only do salts help the body maintain a very narrow blood pH range of 7.35 to 7.45 for survival, but as electrolytes, they also regulate osmotic pressure and electrochemical transport of nutrients—and drugs—into the cells.

And speaking of the realm of health, salt intake and weight have been linked, and researchers are discovering that restricting salt intake can lead to an increase in insulin, a hormone that promotes fat storage, cascading into insulin resistance, which is a major contributor to metabolic syndrome, type 2 diabetes, and heart disease. Even on a micro level, the electrolytes in salt are vital.

In economic terms, as the life force of the economy for time out of mind, salt gave us the word 'salary'. What if this basic combination was not only an elixir of life but also the hottest commodity on Earth? As it once was.

The Salt Dialectic

The struggle for global power and wealth wasn't always centered on oil. For much of human history, it was all about salt. For millennia, salt's pursuit was the very name of a global struggle for power, its mastery a feat of economic and political genius. The fact is, salt was more important for longer throughout human history than oil.

The importance of oil is a relatively recent

phenomenon, dating back only to the late 19th century with the rise of the internal combustion engine and industrialization. While it has been the single most important commodity for the last 150 years, its reign is brief when compared to salt. Oil has only been a major driver of geopolitics for a little over a century, whereas salt's essential role in commerce, political power, and human survival as a life-and-death commodity spans millennia, from the dawn of civilization until the invention of modern refrigeration.

This simple compound, essential for preserving food and sustaining life, was at the heart of empires, conflicts, and monopolies. Salt was a matter of life and death because it was crucial for food preservation before the age of refrigeration. Because of its direct link to survival, it also served as a key source of government revenue and control, making its scarcity or abundance a political weapon. Where so many modern counterfeits flourish, oil's recent reign seems paltry when compared to salt's enduring legacy as the ultimate commodity.

Now the pendulum of history has swung in the other direction, with the unveiling of a highly anticipated salt application: the sodium-ion battery—a breakthrough that could spell the end for conventional lithium-ion phosphate batteries—by the world's largest battery manufacturer, Contemporary Amperex Technology Co., Limited (CATL), located in Ningde, China. Despite sensational rumors circulating, the facts are dramatic enough: the new battery, named NXTra, means a massive ~65% price reduction from current lithium-ion phosphate batteries, and it is not made from rare alkali metals, but from sodium (Na), found in humble salt. Salt is one of the top ten most abundant elements in the Earth's crust, and is about 500 times more abundant than lithium.

The dictum offered by Villanova is an archaic whisper in the modern world. He shows the arcane substance–salt–elevated from common seasoning to a concentrated key. Its perfect, crystalline masterpiece, unlocked by fire and transformation, is the ultimate puzzle, whose influence suggests that power lies not in the obvious, combustible energy of oil, but in a hidden perfection that defies the very nature of combustion. A mineral is a naturally occurring inorganic solid with a specific crystalline structure, while a salt is a type of chemical compound. However, some salts that occur naturally as solids with crystalline structure qualify as minerals. Many important minerals, like calcium phosphate, potassium, and magnesium, are composed of salts. Tradition has it that the one who grasps this subtle balance, this profound chemical art of salts, holds the secret to an eternal truth.

Even in Western medicine, over 50% of all active pharmaceutical ingredients in commercially available drug products are formulated as pharmaceutical salts. From the vital electrolytes that drive cellular function to the crystalline formations that stabilize geological structures, mineral salts are the unassuming chemical bedrock of all life and matter. Salts offer a profound complexity that compels ongoing study, even as researchers continue to uncover the mysteries of salt.

Salt's last peak in popularity was perhaps in the 1870s, when German physician Dr. Wilhelm Heinrich Schüßler pioneered the field of Biochemistry by analyzing the ash residue of human remains to isolate the twelve fundamental inorganic mineral salts essential to the body. This revolutionary work established his theory that imbalances in these 'tissue salts' were the root cause of disease, laying the historical foundation for Cell Salt Therapy as an alternative medical system.

Ancestral Knowledge

Because of industrialization, our ancestors

had knowledge of salt that we lack. Even into the 1950s, people knew much more about the powers of salt. Take, for example, the range of now forgotten recipes they would use it in, such as a precursor to the soda we now buy readymade in a can:

The fundamental concept of using lemon juice (acid) and baking soda (base) to create an instant, fizzy, homemade soda was a common and popular method for making 'fizzy lemonade' or 'quick soda' for decades, into the 1950s.

This lemon soda is good after meals and before a workout to reduce acidity in the muscles, whereas it should be avoided before meals, as it could lower stomach acid necessary for digestion. The chemical reaction between the citric acid in the lemon juice and the sodium bicarbonate (baking soda) releases carbon dioxide gas (CO_2), which creates the fizz.

Dakar Lemon Fizz Soda

Ingredients for one serving

1 lemon
1/4 to 1/2 teaspoon sugar (or to taste)
1/4 to 1/2 teaspoon baking soda (Sodium Bicarbonate)
1 cup cold water
Ice (optional)

Prepare the Lemon: Squeeze all the juice from the lemon into a tall glass.
Add Sweetener: Stir in the sugar until it is mostly dissolved in the lemon juice. Taste and adjust the sweetness.

Add Water: Pour in about 1 cup of cold water and stir to combine.

Add the Fizz: Add the baking soda to the glass.

Stir and Serve: Immediately give it a quick stir. The mixture will fizz up instantly as the carbon dioxide gas is released.

Serve immediately while it's still fizzy. You can add ice if desired.

Tip: Start with 1/4 teaspoon of baking soda. If you want a stronger fizz, you can try up to 1/2 teaspoon, but be aware that more baking soda will increase the salty flavor. You may need to add more sugar to balance the flavor.

2

Salt: The Lifeblood of Empires

"The man who owns the salt,
owns the world."
— Roman Cassiodorus (485-585 AD),
Roman Statesman & Writer

During the thousands of years while salt has been critical to human survival and civilization, its importance was profoundly reflected in the historical power of salt monopolies. This long, historical period passed, much like salt trickling from one hourglass chamber to another.

Humans have relied on salt for as long as they have existed, with the earliest evidence of its use as a commodity dating back to at least 6050 BCE in Romania, and around 6000 BCE in China. This makes salt one of the oldest and most consistently used commodities in human history. The relationship between humans and salt has been pivotal to the development of civilization for ages.

Even Jules Verne describes the detailed process for extracting salt from seawater in *The Mysterious Island.* In the novel, the castaways, led by the engineer Cyrus Smith, need to extract salt to preserve food. They accomplish this by:

- Digging a basin in the sand at a specific point on the beach where the tide won't wash it away.
- Lining the basin with clay to make it waterproof.

- Allowing seawater to fill the basin during high tide.
- Waiting for the sun to evaporate the water, leaving behind a layer of crystallized salt.

A practical, resourceful method, and classic example of the ingenuity that defines the characters' struggle for survival on the island.

Before refrigeration, salt was the only reliable way to preserve food, particularly meat and fish. Control over salt supplies meant control over food, which was essential for feeding armies, sustaining populations during the winter, and enabling long-distance trade.

People couldn't get fresh meat regularly in sixth-century England because of the lack of modern refrigeration and food preservation methods. They had to rely on salt because it was the most common and effective way to preserve meat for long-term storage.

In Mark Twain's *A Connecticut Yankee in King Arthur's Court,* Hank Morgan read off the Dowley family's bill, emphasizing the high cost of goods in their time, and by extension, the benefits of his modern economic system.

> ...Those three amazed men listened, and serene waves of satisfaction rolled over my soul and alternate waves of terror and admiration surged over Marco's:
>
> 2 pounds salt 200. 8
> dozen pints beer, in the wood . . . . . 800.
> 3 bushels wheat . . . .
>
> "And now consider what is come to pass," said he, impressively. "Two times in every month there is fresh meat upon my table." He made a pause here, to let that fact sink home, then added–"and eight times salt meat."
>
> "It is even true," said the wheelwright, with bated breath.

That ratio was a change for the blacksmith Dowley and his family. Observing the living conditions of a common family in King Arthur's time, Hank is points out how the blacksmith's newfound prosperity, thanks to the spread of modern ideas, has improved his diet: The ratio of having two fresh meats to eight salted meats in a month was an immense luxury for the Dowley family. It represents a significant improvement in their standard of living, contrasting their current situation with the extreme poverty and reliance solely on preserved foods that was the norm before Hank's influence.

In the era of King Arthur, fresh meat was scarce. The concept of keeping meat fresh for more than a few days was virtually nonexistent for the general population. Without electricity or iceboxes, fresh meat was a luxury only available immediately after an animal was slaughtered. Most livestock were butchered in the late autumn to avoid feeding them through the harsh winter. This meant that fresh meat was a seasonal food, leaving a long period of the year with no fresh supply.

The Role of Salt as a Preservative

To solve the problem of year-round food supply, people turned to preservation methods. Salt was the most vital of these. The process of curing meat with salt works by drawing out moisture through osmosis. Since bacteria and other microorganisms need moisture to grow and cause spoilage, removing the water effectively prevents the meat from rotting.

This salted, or 'cured', meat could be stored for months on end, providing a crucial source of protein during the winter and other times when fresh meat was unavailable. For this reason, salt was a highly valuable commodity, as Hank Morgan's observation in the book highlights.

Salt in Currency and Trade

Throughout history, salt's ability to preserve food, especially meat and fish, was revolutionary, enabling long-distance travel, military campaigns, and trade. Control over salt routes led to empires, and salt was so valuable that it was sometimes used as currency. The term 'salary' comes from the Latin word *salarium,* which refers to the money paid to Roman soldiers to buy salt. Salt trade routes crisscrossed continents, shaping empires and driving economies. This deep-rooted historical value underscores its status as an elixir of life.

Political power consolidated around salt. Governments and rulers often created salt monopolies to generate revenue and exert control over their subjects. As seen in ancient China's state-run monopoly and the French *gabelle,* controlling salt was a powerful tool for maintaining authority and funding military campaigns.

3

Salt Wars

"Salt is not just a commodity; it is a fundamental right of every human being."
— Mahatma Gandhi, explaining his motivation for the Salt March

---

One might ask, if salt was a hot commodity longer than oil, why didn't people fight over it? The answer: They did. There were many wars over salt. Skirmishes and battles were often preceded or followed by great salt monopolies. In fact, the current geopolitical struggle between the United States and China over rare earth minerals is powerfully illuminated through the historical lens of salt monopolies and salt wars.

Just as salt was a hot commodity longer than oil, vital for basic human life and commerce, rare earth minerals are now the hot commodity essential for all modern technology, from smartphones and electric vehicles (EVs) to advanced military weaponry. The historical salt monopolies, like the infamous *gabelle* salt tax in pre-revolutionary France, were not about market competition but the state's absolute control over a vital commodity for the purpose of immense revenue. This system perfectly mirrors China's current strategy: it's not simply an economic advantage, but a conscious, decades-long policy of state control over the entire industrial ecosystem, encompassing mining, refining, processing, and magnet production, allowing Beijing to hold a near-monopoly on a material the rest of the world desperately needs.

Similarly, the *gabelle* is one of the most famous historical examples of a state-controlled salt monopoly exploiting taxation.

This system, in place for centuries, was a form of price-fixing and monopoly control. The government dictated where salt could be purchased and at what price, making it a source of immense revenue. This system was not about competition among producers but about the state's absolute control over a vital commodity. The *gabelle* led to widespread smuggling, rebellion, and deep resentment among the populace, highlighting how monopoly tactics, when implemented on a national scale, can have devastating social and political consequences. It was one of many 'salt wars'.

More than a Commodity

A salt war occurred in the Ottoman Empire. In the late 19th and early 20th centuries, the Ottoman Public Debt Administration (OPDA), a body controlled by European creditors, took over the salt monopoly from the Ottoman state. The OPDA's goal was to maximize profits to repay the empire's debts.

This led to a harsh crackdown on illegal salt production and smuggling. The OPDA hired its own armed guards, known as 'salt gendarmerie', who would raid villages and confiscate salt. The tactics were brutal and often led to violence, as local communities who relied on small-scale salt production for survival were crushed by a foreign-controlled monopoly.

Salt and the Siege of Venice

Salt brought the Rise of Venice. The great maritime power of Venice in the Middle Ages was built on salt. The city-state didn't just sell salt; it controlled the entire supply chain. Its powerful navy enforced a strict monopoly in the Adriatic Sea, forcing other cities to buy Venetian salt at exorbitant prices. When rivals like Genoa tried to compete, Venice would engage in 'salt wars', sinking ships and blockading ports to maintain its dominance. Venice's control over the salt trade was so absolute that it was central to the city's wealth and political influence for centuries.

For the mercantile Venetians, controlling salt meant controlling life itself for many in the region. Without salt, food rotted, and armies couldn't be provisioned for long campaigns. Venice's ruthless pursuit of a monopoly wasn't about a simple commodity; it was about strategic power. By starving their rivals of salt, they could cripple economies, incite internal unrest, and force city-states to submit to their will. A salt blockade was as effective, and as deadly, as a military siege.

In medieval times, the powerful city-state of Venice maintained a strict monopoly on the salt trade in the Adriatic Sea. Salt was a crucial commodity for preserving food, and Venice used its naval power to control and enforce its monopoly. The city-state forbade its subjects from purchasing salt from any source other than its own government-run depots. It engaged in salt wars with the rival city-state of Genoa, which also sought to control the lucrative trade routes. Venice's tactics included blockading rival ports and sinking ships that were carrying unauthorized salt. This historical example shows how a monopoly could be enforced through military and political power rather than just economic manipulation.

Despite its importance, trade in brine, very salty water, often a byproduct of other processes, was slow to modernize. For centuries it presented the spectacle of a technological desert, a vast indifference to innovation driven by monopolistic greed. Modernizing the ancient, open-pan process became a major stumbling block. Despite three hundred years of patents that might have offered superior improvements, every attempt to refine the process was met with a resolute, active opposition from the powerful, established proprietors, who saw stagnation not as a failing, but as their most dependable crutch. Salt began its final decline during the industrial revolution. According to Albert Frederick Calvert's *Salt and the Salt Industry:*

> Between 1846 and 1880, the trade was being continually reorganized for offensive and defensive commercial purposes, but, in 1881, it was admitted that, in spite of all attempts to encourage a better feeling among the leading manufacturers, "the spirit of envy, hate, malice, and all uncharitableness, which has so long been the bane of the salt trade, has again become rampant," with the result that the price of common salt–4s., less the brokers' discount of 5 per cent.–was the lowest that it had touched since the American Civil War. Two years later, it was declared that the trade, instead of being ruled by common sense and business experience, was being ruined by personal animosities and trade jealousies.

These salt industry owners maintained their crude, wasteful methods because the resulting monopoly profits were extraordinarily large, giving them no financial incentive to embrace change or allow competitors who used new, efficient plants to enter the market. Their opposition was so fierce that they not only crushed external innovators, but their internal trade jealousies and animosities led to continuous, ruinous price-cutting wars, forcing them into successive, ultimately unsuccessful combinations to regulate stocks and prices.

Oil: The Fuel of the Modernist World

The dominance of oil, while powerful, is a modern story. It didn't begin until the mid-19th century. Then, its rise to global importance was cemented in the 20th century, during the Industrial Revolution. Oil's true value was unlocked as a fuel source for factories, transportation (cars, ships, planes), and power generation. This fueled the industrial and technological advancements that define the modern world. The rise of oil brought Geopolitical Power shifts. Where so many commodities—salt, coal, and the like—were democratically scattered across the map, oil's dominance is a concentrated inequity by twentieth-century standards, a seeming display of nature's favoritism. This has made oil, not salt, the central factor in international relations, military conflicts, and global politics for the last century.

While oil's concentrated power is a seeming display of nature's favoritism, its genesis is a far more retrograde karmic balancing act–a victory of the small over the large. Unlike salt, which can be found in many parts of the world, major oil reserves are concentrated in specific desert regions that once flourished as centers of life. This liquid paragon doesn't come from the magnificent, literal fossils of land animals like dinosaurs, which were too quickly devalued by decay. No, oil is a subtle accumulation of microscopic marine organisms–algae and plankton–that sank to an anoxic, oxygen-poor grave. This organic sludge was buried under immense pressure and heat over millions of years, a profound, slow-motion transformation where the large molecules of organic matter crack and convert, through the 'Oil Window', into the smaller, pure liquid hydrocarbons we call crude. Oil's dominance, then, is the flipside of the ultimate devaluation of the barely visible, its power a tribute to the invisible, ancient death of the infinizillion plankton from which it originates.

The destructive consequences of the salt monopolies, which led to widespread smuggling, rebellion, and the harsh tactics of the 'salt gendarmerie' in the Ottoman Empire, serve as a dramatic historical warning for the current supply chain crisis. The Ottoman Public Debt Administration (OPDA), controlled by European creditors, crushed local salt producers to maximize profit for debt repayment, highlighting the brutal power of a foreign-controlled monopoly over a necessary resource. Today,

China has responded to U.S. weaponization of the dollar by weaponizing its control over rare earth minerals in a similar show of force. The recent sweeping export curbs on rare earth, lithium-ion batteries, and graphite are Beijing's modern-day salt gendarmerie, a tactical move to enforce control and maximize leverage in the U.S.'s trade war.

History Repeats

The key lesson from the salt wars is that controlling the choke point is the ultimate source of power. In the 19th-century Ottoman Empire, the choke point was the salt production itself; today, the choke point is the processing capacity, as China controls over 90% of global rare earth refining. This means that even when allies like the United States or Australia extract raw rare earth, they must still send them to China for the critical refinement stage—the equivalent of the Ottoman salt smugglers being forced to use the OPDA's channels. This dependency has staggering implications, as nearly 78% of the 1,900 US military weapon systems rely on Chinese-sourced minerals, a direct vulnerability that allows China to exert power over the US military-industrial complex akin to the devastating social control the gabelle had over the French populace. The struggle is no longer about economics; in both the past struggle for the 'white gold' and the current race for the 'rock-to-rocket' minerals, it is fundamentally a fight for national survival and ultimate power.

4

What is Salt?

"Take it with a grain of salt."
—Anon

From a modern perspective, it's not easy to see what all the fuss over salt was about. What is it about salt that makes this common commodity the center of so much gravity?

It begins with the sea. In the scorching sun, sea water gives up its essence, evaporating until only the white, crystalline bones of its former self are left behind. This residue was the first substance humanity named 'salt'. Originally, the term 'salt' only referred to the white residue left behind after seawater evaporated. But the word's meaning, like a great wave, kept expanding. Over time, its meaning expanded to include all the other minerals dissolved in the sea.

Later still, chemists gave the word a much broader definition. It was stretched to include all the other minerals pulled from the sea. And then, the scientists took hold of it, pushing its definition even further. They saw the same fundamental combination in countless other substances.

In nature, when the acid hydrochloric acid (HCl) reacts with the base sodium hydroxide (NaOH), it produces the salt sodium chloride (NaCl)–common table salt–along with water. This attraction of opposites is memorialized in salt crystals, or as the century philosopher put it,

> "If I myself am a grain of the saving
> salt which maketh everything in the
> confection-bowl mix well: – For there
> is a salt which uniteth good with evil;

> and even the evilest is worthy, as spicing and as final over-foaming…"
>
> – Friedrich Neitzsche,
> *Thus Spake Zarathustra: A Book for All and None*

In the language of chemistry, 'salt' became a kind of cosmic family name. The simplest member of this family is what we now call common salt. Its secret is a perfect, atomic embrace. A single, unstable atom of the metal sodium finds its one true partner in an atom of the poisonous gas chlorine. They merge, their unstable natures cancelling each other out in a perfect, stable union. It's a desperate attraction, a powerful, quiet bond that creates the very substance of life and taste: NaCl.

Facets of Salt

But the world is full of much more than chemists. Salt has meant many things to many kinds of people. New words are constantly created, and the etymological relationships can be complex. Many common English words are directly derived from salt or its proto-Indo-European root 'sal-'. Another Latin root is *Salus*, the Roman goddess of safety, well-being, health, and prosperity. Words like 'salient' and 'result' are derived from the Latin verb *salire*, meaning 'to leap' or 'to jump'. The words 'salire' and 'salt' share a similar sound and the Latin root 'sal-'.

- **Salire (to leap):** This Latin verb is the root of many English words related to jumping, springing, or projecting.
    - **Salient:** 'Leaping out' or prominent.
    - **Assault:** A 'leap at' an enemy.
    - **Exult:** To 'leap up' with joy.
    - **Result:** An outcome that 'leaps back' from a cause.
    - **Somersault:** A leap or flip.
    - **Resilient:** 'Leaping back' or springing back into shape.
- **Salt:** The English word 'salt' comes from the Proto-Germanic root *saltom, which is related to the Latin word *sāl.* This root refers specifically to the compound.

The most famous examples come from the Latin word sal, which was not just a seasoning but a precious commodity used for trade and payment. This historical significance is reflected in the word salary.

'Salty' describes a person who is angry, bitter, or irritated, often over a minor disappointment or perceived slight. While it has recently surged in popularity in online gaming communities, its roots are much older. One of the most widely accepted theories for the origin of this figurative meaning is a direct connection to the taste of tears. A 'salty' person is someone who is metaphorically or literally shedding bitter, salty tears of frustration or sadness. This association dates back at least to the 1930s. Science has proven the relationship between salt and sex drive with the discovery that sodium, the very essence of salt, becomes a crucial element in regulating the fluid and electrolyte balance that is, in turn, heavily managed by the adrenals. These glands are also the masters of the steroid organs, the source of our sex hormones. Thus, salt's simple chemistry is intimately, if subtly, entangled with the body's most potent players, its naive presence essential to the functioning of those magnificent, concentrated compounds that orchestrate the human drive.

Another theory, which might have contributed to the term's meaning, comes from naval slang. 'Old salts' were seasoned sailors who had spent a lifetime at sea, enduring harsh conditions and the constant exposure to saltwater. This lifestyle was thought to make them tough, but also grumpy and irritable, leading to the phrase 'jump salty' to describe someone suddenly becoming enraged.

'A Salt'

In *Moby Dick*, Herman Melville explores the various facets of salt, from the literal salt spray of the sea to its metaphorical meaning in phrases like 'a kind of man'.

Herman Melville displays the aspects of salt in *Moby Dick*, shedding light through sprays of saltwater down upon the metaphorical meaning for a certain kind of man. As the old salt put it:

> I never go as a passenger; nor, though I am something of a salt, do I ever go to sea as a Commodore, or a Captain, or a Cook. I abandon the glory and distinction of such offices to those who like them. For my part, I abominate all honorable respectable toils,

trials, and tribulations of every kind whatsoever. It is quite as much as I can do to take care of myself, without taking care of ships, barques, brigs, schooners, and what not. And as for going as cook...yet, somehow, I never fancied broiling fowls;–though once broiled, judiciously buttered, and judgmatically salted and peppered, there is no one who will speak more respectfully, not to say reverentially, of a broiled fowl than I will. It is out of the idolatrous dotings of the old Egyptians upon broiled ibis and roasted river horse, that you see the mummies of those creatures in their huge bake-houses the pyramids.

No, when I go to sea, I go as a simple sailor, right before the mast, plumb down into the forecastle, aloft there to the royal mast-head. True, they rather order me about some, and make me jump from spar to spar, like a grasshopper in a May meadow. And at first, this sort of thing is unpleasant enough. It touches one's sense of honor, particularly if you come of an old established family in the land, the Van Rensselaers, or Randolphs, or Hardicanutes. And more than all, if just previous to putting your hand into the tar-pot, you have been lording it as a country schoolmaster, making the tallest boys stand in awe of you. The transition is a keen one, I assure you, from a schoolmaster to a sailor, and requires a strong decoction of Seneca and the Stoics to enable you to grin and bear it. But even this wears off in time.

What of it, if some old hunks of a sea-captain orders me to get a broom and sweep down the decks? What does that indignity amount to, weighed, I mean, in the scales of the New Testament? Do you think the archangel Gabriel thinks anything the less of me, because I promptly and respectfully obey that old hunks in that particular instance? Who ain't a slave? Tell me that. Well, then, however the old sea-captains may order me about—however they may thump and punch me about, I have the satisfaction of knowing that it is all right; that everybody else is one way or other served in much the same way—either in a physical or metaphysical point of view, that is; and so the universal thump is passed round, and all hands should rub each other's shoulder-blades, and be content.

The English language has defined itself through the words derived from 'salt'. The importance of salt throughout history, from a precious commodity to a form of payment, has imprinted our language.

Many common words we use today are directly derived from the Latin root *sal* or related terms:

Salary

The most famous example, the word *salary* comes from the Latin *salarium,* which was an allowance given to Roman soldiers for the purchase of salt. This highlights salt's immense value and its function as a form of currency. The expression 'worth one's salt' also stems from this historical context, meaning someone is competent and capable enough to earn their keep.

Salad

The word 'salad' comes from the Latin *salsus,* meaning 'salted'. Ancient Romans enjoyed their leafy greens seasoned with a dressing of oil, vinegar, and, most importantly, salt. So, a 'salad' literally means 'salted vegetables'.

Sausage

The term 'sausage' also traces its origins to *salsus.* The word refers to meat that has been seasoned with salt and other spices and then cured or preserved.

Sauce

Much like 'salad' and 'sausage', the word 'sauce' comes from *salsus.* Salt was a fundamental and essential component of ancient sauces and condiments, used to both enhance flavor and preserve ingredients.

Saline

This adjective describes something containing salt or having the characteristics of salt. It comes from the Latin *salinus,* which means 'of salt'.

Silt

This word for fine sediment or mud is

believed to be related to the Scandinavian term *sylt*, which referred to a 'salt marsh' or 'brine'. This connection highlights the geological deposits left behind by saltwater.

Culinary and Food-Related Words

Salami: From the Italian *salame*, which is related to *sale* (salt).

Salsify: A vegetable whose name comes from the Old French *salsifi*, with a possible link to salt due to its taste or preparation.

Chemical and Scientific Words

Salinity: The amount of salt dissolved in a body of water.

Halite: (NaCl), or table salt, is the most famous kind of salt; the mineral compound form of salt (sodium chloride), from the Greek *háls* (salt).

Halogen: The group of chemical elements including fluorine, chlorine, bromine, and iodine, all of which form salts when they react with metals.

Haloid: A term used to describe a salt-like compound.

Desalination: The process of removing salt from water.

Geographical and Geological Words

Saltern: A place where salt is made by evaporation.

Salt marsh: A marshy area that is regularly flooded by seawater.

Historical and Nautical Terms

Saltcellar: A container for holding salt.

Salting: The process of preserving food with salt.

Salt away: To save or hoard money, often in secret, like hoarding a valuable commodity.

Old salt: A seasoned sailor, referencing their exposure to the sea.

Worth one's salt: To be competent or deserving of one's pay.

Salt of the earth: A biblical phrase describing a person of great value and integrity.

Other Derivatives and Compounds

Brine: A highly concentrated solution of salt in water.

Presalt: A term used in geology to describe rock formations that predate a thick layer of salt.

Undersalt: To use too little salt.

Oversalt: To use too much salt.

Rock salt: The common name for the mineral halite.

Salt shaker: A container with small holes for sprinkling salt.

An Acid and a Base

Remember that in chemistry, a salt is any compound formed when a base and an acid combine. Our familiar table salt, or sodium chloride (NaCl), is a perfect and simple example of this. Its molecule consists of just two parts: one atom of sodium, a highly reactive metal, and one atom of chlorine, a toxic gas. Alone, they are dangerous; together, they form a stable and essential compound.

This perfect one-to-one pairing is due to their monovalent nature. This means each atom can bond with or replace a single hydrogen atom in a compound. This simple atomic partnership is the basis for all the diverse substances we now classify as salts.

Now modern chemistry describes salts' properties, heretofore seen in terms of ancient philosophies, for example of yin and yang. Science has demonstrated how an electron-deficient acid and an electron-excessive base combine to create a balanced, stable salt crystal. This ancient quest for balance was finally given a name in 1744, when a member of the French Royal Academy of Sciences, Guillaume Francois Rouelle, provided a definition that endures to this day. He declared that a salt was any substance created by the reaction of an acid and a base. For a long time, these two substances were little more than curious opposites–acids were sour and dissolved metal, while bases felt soapy to the touch. But what was it that drew them together?

The answer lay in a profound, almost mystical affinity. Nature, in its pursuit of perfection, had designed acids and bases to be a natural pair. In common table salt, the electron-donating base is sodium, and the electron-hungry acid is chloride. But the process of forming common table salt (NaCl) is an electron transfer reaction, which is properly categorized as a Redox (Reduction-Oxidation) reaction, not a traditional acid-base reaction. In this process, the neutral sodium atom (Na) is oxidized as it loses a single electron to become a positive ion (Na^+), making it the reducing agent. Conversely, the chlorine atom (Cl) is reduced as it gains that electron to become a negative chloride ion (Cl^-), making it the oxidizing agent.

A more sophisticated Lewis acid-base theory deals with the donation or acceptance of an electron pair to form a covalent bond, whereas the formation of NaCl involves the transfer of only a single electron to form an ionic bond. The resulting compound, NaCl, is the product of a strong acid (HCl) and a strong base (NaOH), making it a neutral salt that exhibits neither acidic nor basic properties when dissolved in water.

Pure salt (sodium chloride) isn't deliquescent, meaning it doesn't melt by absorbing moisture from the air. However, raw salt and even commercial salt often do this because they contain tiny amounts of other substances, especially magnesium chloride, which is highly deliquescent. Even the most refined table salt is still hygroscopic, meaning its crystals can absorb a small amount of moisture (up to 0.6%) from a humid environment.

Other kinds of salts and Their Components

The world is full of countless types of salts, both for culinary and industrial uses, as well as geological minerals, since a salt is formed anytime an acid and a base react.

The simple principle of electron donation and acceptance, the very act that creates a substance as ordinary as table salt (sodium chloride), is a universal rule in the proliferation of all salts. The electron donor, always a positively charged ion, or cation, finds its perfect match in the electron recipient, a negatively charged ion, or anion. Where so many acids and bases must interact, this foundational rule–donation and acceptance–creates nature's masterpiece: a neutral, stable, and essential salt.

In the salt industry, workers labor to coax a

potent complex of over 70 trace minerals from brine, including life-sustaining magnesium, vital potassium, and even trace amounts of lithium. This concentrated harvest is the foundational ingredient for modern healing: it is bottled as electrolyte concentrates to rehydrate the exhausted, and formulated into supplements promising relief for the strained mind, strong bone, and weary muscle. Yet, the uses extend beyond wellness. Magnesium is wrenched from the brine to produce the ultralight metal titanium, the same material that becomes the unyielding skeleton of an artificial hip, a miraculous replacement that grants mobility to the crippled. Salt deposits, therefore, are no inert flats, but immense, silent quarries fueling both the subtle chemistry of the body and the enduring art of modern medical engineering.

Some of the most well-known salts:

- Sodium Chloride (NaCl): Also known as table salt or common salt.
  - o Electron Donor (Cation): Sodium (Na+)
  - o Electron Recipient (Anion): Chloride (Cl−)
- Potassium Chloride (KCl): A common salt substitute, often used for flavoring or as a fertilizer. Found in many salt substitutes, potassium chloride is made from the reaction of potassium hydroxide (a base) and hydrochloric acid. In this salt, the potassium ion (K+) is the electron donor, and the chloride ion (Cl−) is the electron recipient.
  - o Electron Donor (Cation): Potassium (K+)
  - o Electron Recipient (Anion): Chloride (Cl−)
- Magnesium Sulfate (MgSO4): Better known as Epsom salt, it is used for muscle relaxation and in gardening. This compound is formed from magnesium hydroxide (a base) and sulfuric acid. The magnesium ion (Mg2+) donates two electrons, and the **sulfate ion (SO42−) ** accepts them.
  - o Electron Donor (Cation): Magnesium (Mg2+)
  - o Electron Recipient (Anion): Sulfate (SO42−)
- Calcium Carbonate (CaCO3): A key component of limestone and chalk, it is also used as a dietary supplement. It's created by the reaction between calcium hydroxide (a base) and carbonic acid. The calcium ion (Ca2+) is the electron donor, and the **carbonate ion (CO32−) ** is the electron recipient.
  - o Electron Donor (Cation): Calcium (Ca2+)
  - o Electron Recipient (Anion): Carbonate (CO32−)

- Ammonium Nitrate (NH_4NO_3): A crucial ingredient in many fertilizers and explosives, this salt is formed from the reaction of ammonia (a base) and nitric acid. The ammonium ion (NH_4^+) donates an electron, and the nitrate ion (NO_3^-) accepts it.
  - Electron Donor (Cation): Ammonium (NH_4^+)
  - Electron Recipient (Anion): Nitrate (NO_3^-)
- Sodium Bicarbonate ($NaHCO_3$): More commonly known as baking soda, it is a leavening agent in baking and an antacid.
  - Electron Donor (Cation): Sodium (Na^+)
  - Electron Recipient (Anion): Bicarbonate (HCO_3^-)
- Potassium Iodide (KI): Used in medicine and as a source of iodine in iodized salt.
  - Electron Donor (Cation): Potassium (K^+)
  - Electron Recipient (Anion): Iodide (I^-)
- Lithium Carbonate (Li_2CO_3): Used as a mood stabilizer in psychiatric medicine.
  - Electron Donor (Cation): Lithium (Li^+)
  - Electron Recipient (Anion): Carbonate (CO_3^{2-})

5

The Salt Monopolies

"Salt is a great political matter of the state, serving to enrich the nation, strengthen the army, and benefit the people by fostering prosperity."
— Sima Qian in his Shiji (Records of the Grand Historian), from the Chinese: "鹽者，國家之大政也，所以富國強，利民興利者也. " (Pinyin: "Yán zhě, guójiā zhī dàzhèng yě, suǒyǐ fù guó qiáng bīng, lì mín xīng lì zhě yě.")

Salt's critical, non-negotiable role in daily life made it an easy and lucrative target for governments seeking revenue. Consequently, the power to control and tax this basic substance became a lever of political oppression, leading to centuries of widespread resistance. Taxes on salt–a requirement for life itself–have fueled revolutions. In France, the government salt monopoly, a simple act of government control, was a masterpiece of tyranny, a symbol of royal rule and the exploitation of the populace. The *gabelle,* or tax on salt, a kind of primitive necessity, became a naive act that would bring the nation to its knees.

In pre-revolutionary France, the government's monopoly on salt was a key source of revenue and a symbol of royal tyranny, and a tax on survival. This humble commodity, salt, was the only reliable way to preserve meat and fish through the winter, preventing starvation. The state's monopoly and mandatory purchase laws meant that poor families had to choose between paying an exorbitant tax on their salt and risking their food supply rotting. Smuggling salt, a crime punishable by death or a long, bleak sentence in the galleys, was a desperate, almost naive, bid for survival for those trying to feed their families. For them, salt wasn't a seasoning–it was the thin line between a full belly and the grim specter of famine. The *gabelle* salt tax, was a crushing burden on the poor.

The state dictated where people could buy salt, how much they had to buy, and at what price. This system of control led to widespread smuggling and rebellion, becoming one of the most hated and deeply resented taxes in French history. The anger over the *gabelle* was a significant contributing factor to the French Revolution, showing how control over a common commodity could ignite a political firestorm.

Ruthless competition also pervaded American salt fields. In the early American salt industry, particularly in the Kanawha Valley of West Virginia and in Syracuse, New York, the struggle for dominance was fierce. During the 19th century, salt producers formed associations and cartels to control prices and crush competition. In the Kanawha Valley, for example, salt makers created the Kanawha Salt Company, a powerful combination that aimed to regulate output and maintain high prices. They would sometimes deliberately flood the market with cheap salt to bankrupt smaller, independent producers, a classic 'war of extermination' tactic. This led to a boom-and-bust cycle, as described in the passage, where profits were often 'thrown away' in an effort to secure market control. The 'Salt City' of Syracuse saw similar struggles, with different salt companies vying for control over the valuable brine springs.

Another example of a historical salt monopoly and the ruthless tactics used to maintain it was the Liverpool Salt Association in 1888. The Liverpool Salt Association was formed by Cheshire salt manufacturers to combat the problem of overproduction and falling prices. Its aim was to restore profitability by artificially restricting output and fixing prices. The association operated as a cartel, with members agreeing to reduce production by 50% and sell their salt through a central office at a minimum price. This was a classic example of a price-fixing scheme. The Liverpool Salt Association's tactics were a direct response to the 'war of extermination', showing how competitors, after years of mutual destruction, would reluctantly band together to create a monopoly.

The Liverpool Salt Association's formation was a reluctant feat, a kind of naive retreat from the 'war of extermination' that had defined the industry. Where so many ruthless tactics flourished, carrying off a masterpiece of a monopoly seemed a desperate last resort. This banding together–this quiet collusion among rivals–was not an act of camaraderie or common interest. Rather, it was a direct and joyless response to a long-held, self-destructive pleasure: the unlimited proliferation, and devaluation, of competition itself. After years of mutual ruin, the establishment of the cartel became a promise of stability and profit that, in its very essence, marked the end of a more rugged, if ruinous, form of commerce.

The *gabelle,* an oft-repeated dynamic, illustrates how a government's weaponization of basic human necessities for profit and control, degrades a means of survival into a symbol of oppression. This created an unjust exercise in cruelty, where the poor were punished for the simple act of trying to survive. The lessons learned from this historical episode recur throughout history, and can be applied to modern contexts.

The Salt Barons of China

In Imperial China, the salt monopoly was a matter of life and death for the populace and the state. For the common person, an increase in the price of salt, manipulated by the barons, meant the inability to preserve food, leading to malnutrition and disease. For the government, the revenue from the salt monopoly was the 'greatest treasure in the world', funding the military and state functions. The barons' power was so immense, they could threaten the very stability of the empire if their privileges were challenged. The salt trade, therefore, was a delicate balance: a source of life for the people and a lifeblood for the state, both of which were held in the hands of a few powerful men.

In Imperial China, the salt trade was a state-run monopoly for over 2,000 years, making it a cornerstone of government finances. The Chinese state held a tight grip on the salt industry, not just as a source of revenue, but as a central pillar of the government's finances, and its imperial authority. The emperor granted exclusive licenses to wealthy families and powerful merchant guilds, effectively creating a class of 'salt barons' who operated with the full backing of the state. These merchants wielded immense power, manipulating prices and supply to eliminate competition. Their wealth was so vast, they could fund military campaigns and influence imperial policy. These licensed salt merchants became an elite class, accumulating staggering fortunes that rivaled, and sometimes even surpassed, that of the imperial court itself. Their wealth was built on two key factors: control and corruption.

The salt merchants operated with a ruthless efficiency, engaging in price and supply manipulation. They could hoard salt to drive up prices, restrict supply to create artificial scarcity, and flood a market to bankrupt any small, independent producers. Their goal wasn't just profit; it was the total elimination of competition. The wealth of these merchants gave them immense political power. They used their fortunes to fund government initiatives, pay for local infrastructure, and even finance military campaigns, essentially becoming private bankers to the state. In return, they enjoyed protection and were often able to influence government policy to serve their own interests. This symbiotic relationship solidified their monopoly.

The Parallels with India's Salt Revolution

This historical context provides a powerful parallel to India's struggle against British colonial rule, culminating in Mahatma Gandhi's Salt Satyagraha (Salt Revolution). In both cases, a government-controlled salt monopoly became a flashpoint for popular dissent.

Prior to the 1882 Act, salt was taxed differently across various provinces. For example, in the late 1870s, a proposal was made to raise the salt tax by 40% in some provinces, which would result in a very considerable increase. A basket of salt that cost three pies to produce was made to cost five annas–a massive markup. (There were 12 pies in an anna, and 16 annas in a rupee.) This demonstrates the immense price-fixing and profit-taking that the tax allowed.

Before Gandhi's Salt March in 1930, the revenue from the salt tax was substantial. It amounted to a significant portion of the total British revenue in India, with some sources claiming it was about £25 million. The tax was especially oppressive for the poor, as it was a flat tax on a necessity. As a result, the tax consumed a disproportionately large percentage of a poor family's income compared to a rich one. This is why Gandhi chose salt as the focal point of his protest–it was a tax on the very air the poor breathed, a symbol of their systematic exploitation. On March 12, 1930, Mahatma Gandhi began the Salt March, a 240-mile walk from his ashram in Ahmedabad to the coastal village of

Dandi. His act of civil disobedience was simple yet powerful: he walked to the water's edge, bent down, and picked up a lump of natural salt, defying the British law that monopolized salt production and taxed it heavily.

The act and its symbolism have been the subject of many books. Gandhi's civil disobedience was not only about the salt; it was a powerful symbolic challenge to British authority. By collecting salt from the Arabian Sea, he was directly violating the 1882 British Salt Act, which forbade Indians from collecting or selling salt and made them pay a tax on a common necessity.

The act was a stroke of strategic genius for several reasons, starting with salt's universality. Salt was a basic need for every Indian, regardless of caste, religion, or class. The salt tax affected everyone equally, making the protest a unifying cause. The law was seen as fundamentally unjust, making the

British government appear oppressive and greedy for taxing a natural resource essential for life. Ghandi took the moral high ground. His act of defiance was nonviolent and easily replicated. This encouraged thousands of Indians to follow his example, leading to mass arrests and widespread civil disobedience across the country.

The Salt March and Gandhi's act of defiance mobilized the Indian population and drew international attention to the injustices of British colonial rule. It was a pivotal moment in India's struggle for independence, demonstrating the power of peaceful resistance against an unjust law. This basic necessity was taxed in both imperial China and colonial India, the government's monopolistic control over salt led to its taxation as a commodity, not a basic human right. The British Salt Act of 1882 gave the British a complete monopoly over salt production and sale in India, imposing a heavy tax that made salt unaffordable for many poor people. Similarly, the Chinese salt barons, with state backing, manipulated prices and supply, forcing the common people to pay exorbitant prices for an essential item.

The Power of the Monopoly

In China, the licensed salt merchants were a formidable alliance of private wealth and state power. They used their influence to fund government projects and even military campaigns in exchange for the right to crush any competition, including small-time smugglers. This mirrored the British in India, who used their colonial power to enforce their monopoly, punishing anyone who produced or sold salt without a license. In both instances, salt became a potent symbol of injustice and oppression. Gandhi's Salt March to Dandi in 1930 was not just an act of defiance against the British tax; it was a powerful, symbolic protest against the entire system of colonial exploitation. By illegally making salt from seawater, Gandhi and his followers highlighted the absurdity of a law that prevented people from collecting a natural resource available on their own shores. The Chinese salt revolts, though often local and decentralized, were similarly driven by popular anger over a government and merchant class that controlled access to a life-sustaining necessity. In both historical narratives, the fight for a grain of salt was, in reality, a fight for freedom and justice.

6

Senegal and Ghana's Salt Empires

"Just because you're rich doesn't mean you should lick salt."
—Edo proverb from Nigeria (a warning against extravagance)

In the sun-scorched lagoons and blinding white pans of West Africa, an ancient struggle unfolds over a mineral more precious than gold to human life: salt. Senegal and Ghana, the sub-region's titans of sodium chloride, are locked in a high-stakes race to control the supply of this essential commodity, with their industries serving as a lifeline–and sometimes a flashpoint–for millions in a 'white gold' rush.

Senegal: The Reign of the Pink Lake

Senegal currently holds the crown, as West Africa's largest salt producer, harvesting over 450,000 tonnes annually. This is a landscape where the industrial might of modern refineries meets the grueling, iconic labor of the past. The image most people hold is that of Lac Rose (Pink Lake), where men scrape the shocking pink crust of the hyper-saline waters into towering, ivory pyramids. This artisanal effort, though small-scale, provides a critical livelihood for thousands, a testament to the sheer human effort required to extract the 'white gold'. Yet, the bulk of Senegal's production and export might flows from large industrial concerns like the National Saloum Salt Company (SNSS). These companies produce refined, high-quality salt, ensuring that Senegal not only feeds its own population but also acts as the primary supplier to landlocked countries in the Sahel, where this sodium chloride (NaCl) is vital for both human health and animal feed.

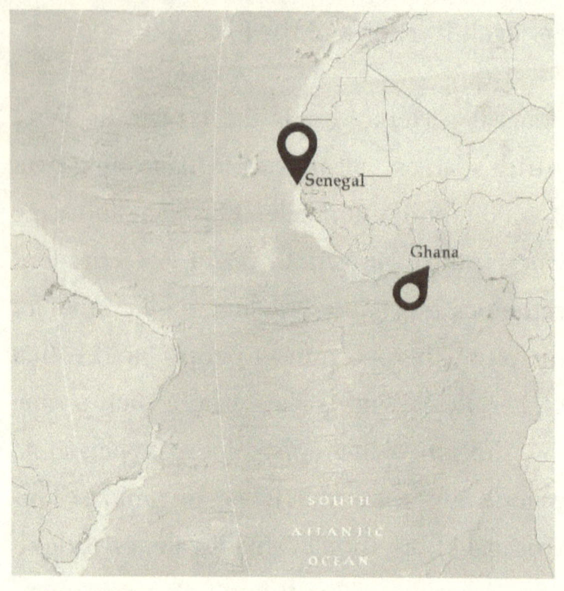

The national imperative, however, is not just about quantity; it's about quality and iodine. In a region plagued by iodine deficiency disorders (IDD), Senegal's policy efforts focus on ensuring that all salt, from the remote artisanal pan to the massive refinery, is adequately iodized. This is a battle fought not with weapons, but with chemistry and logistics, to protect the mental and physical development of millions across West Africa.

Ghana, the Sleeping Giant, Awakens

Further south, Ghana has historically been the region's sleeping giant. With a production of around 250,000 metric tons annually, its output has lagged far behind its potential, estimated to be more than two million tons per year. For decades, the industry was characterized by small-scale solar evaporation and complex land tenure disputes that stifled investment.

Ghana's potential lies in its vast coastal lagoons, especially the legendary Songor Lagoon . This is where the drama of industrialization is currently at its peak. A new era has dawned with massive, focused investment transforming the landscape. The ambitious revamp of the Songor project, led by Electrochem Ghana Limited, aims to catapult the nation's salt production to one million tonnes of high-purity salt annually. This transformation is driven by the soaring global demand for industrial-grade salt used in the petrochemical and chloro-alkali industries, which are crucial for producing chlorine, caustic soda, and other basic chemicals. The vision is to shift Ghana's economic perception of salt from a rudimentary food additive to a strategic, multi-billion-dollar mineral asset.

The Unsung Salts: Beyond NaCl

The battle for West African salt is not only about common table salt, but also about the hidden treasure of mineral salts that make the extraction sites viable.

While NaCl is the star, the evaporation process in the coastal pans and lagoons concentrates a rich cocktail of other elements. For an area like Ghana's Songor Lagoon to reach its true industrial potential, it requires the ability to diversify its product line by separating these other valuable marine chemicals, such as magnesium chloride (MgCl2) and calcium sulphate (CaSO4).

These 'secondary' salts are critical raw materials for pharmaceutical, construction, and agricultural sectors. Senegal and Ghana are thus not just competing to fill a salt shaker; they are competing to establish a regional industrial base.

By maximizing the purity of the common salt and simultaneously harnessing the complex mineral salts, they seek to transition from being mere exporters of a raw commodity to becoming global suppliers for a diversified chemical industry. The 'White Gold Rush' continues, demanding political will, technological investment, and the relentless labor of the people, all focused on transforming saline water into economic destiny.

7

Persian Salt Mountains

"Music is liquid architecture and architecture is frozen music."
—Johann Wolfgang von Goethe

In the parched, sun-scorched landscapes of the Zagros Mountains, Persia, a primal narrative is etched into the very earth–a story of an ancient sea, its slow retreat, and the mineral legacy it left behind. This is not a tale of the living, but of the lifeless, crystalline heart of the region: salt. But here, in this rugged corner of the world, salt is more than a simple seasoning; it is a geological phenomenon, a source of life and conflict, and a silent testament to the immense power of deep time.

The Colorful Persian Salt Mountains

Westerners originally used the term ‘salt’ to describe the residue left after seawater evaporated. Over time, its meaning expanded. First, it came to include other substances dissolved in the sea. Later, chemists broadened the definition even further to describe all compounds formed from a base and an acid. Sodium chloride ($NaCl$), now known as ‘common salt’, is a prime example of a simple chemical salt. Its molecule is made of one sodium atom and one chlorine atom. The precise, scientific classification of salts, however, was already a matter of intuitive knowledge for the people of the Zagros Mountains. In their parched, sun-scorched landscapes, a primal narrative is etched into the very earth—a story of an ancient sea, its slow retreat, and the mineral legacy it left behind. The colorful, striped strata of the Zagros Mountains were a visible testament to this history, and locals learned to distinguish between them, understanding that each color represented a different type of salt, long before chemistry formally defined them.

Salt, in its purest form, is a translucent crystal, perfectly clear and colorless. This pure state, known as anhydrous sodium chloride (NaCl), is a rare find in nature. In most mines, such transparent specimens are considered curiosities. However, in certain deposits, like the famous Wieliczka Salt Mine in Poland, large masses of incredibly pure, transparent salt are common. What most people think of as rock salt is the white, opaque variety. This color comes from microscopic impurities like trapped water, air, and other minerals that break up the light passing through the crystals.

The tale of colored salt is a vibrant one, connecting science to the magnificent salt mountains of Persia. The majestic salt domes of the Zagros Mountains (cover photo) are not just white; they are streaked with an array of colors, a testament to the 'impurities' within the salt. These formations get their brilliant hues from different mineral deposits.

Just like the famous Salt Range in Punjab, the salt mountains in Iran are often colored by the presence of iron oxide, or rust. Depending on the concentration of iron, the salt can range from a soft pink to a deep, dramatic red, making for a truly breathtaking landscape. The color of salt can vary dramatically. Salt from Cheshire, England can be other colors from yellow to reddish-brown, to rare, transparent blocks of deep sapphire blue, sometimes found in the Wieliczka mines. This blue color is a fascinating scientific mystery, attributed by some chemists to the presence of subchloride of sodium and by others to gas inclusions within the crystal's structure. The diverse colors of salt, from the purest transparent crystal to vibrant reds and blues, tell a story written over millions of years, reflecting the unique geological history of each region. The stunning, multi-colored salt mountains of Iran are a powerful example of this natural artistry.

The story begins millions of years ago, in the late Precambrian era, with the demise of the Tethys Sea. As tectonic plates shifted, the sea was caught in a continental squeeze, its waters evaporating under the relentless sun. The minerals dissolved within its depths began to precipitate, layer upon agonizing layer, forming what geologists call evaporite deposits. These layers, compressed over eons, formed vast salt domes, or diapirs, that lie hidden beneath the mountain ranges.

In the heart of the Zagros, these diapirs are not dormant. They are viscoplastic, meaning they flow like a slow, unimaginably thick liquid under immense pressure. The salt, being less dense than the surrounding rock, begins a clandestine journey upward, a geological slow-motion rebellion. It fractures the overlying rock, pushing it aside with inexorable force. This upward migration is a testament to the strange, almost sentient nature of these mineral masses. They pierce the surface in a spectacular display of geological theater, forming salt glaciers that creep down the valleys, and salt caves that are a hauntingly beautiful and surreal landscape.

The most famous of these is Mount Namakdan, or 'saltshaker', on Qeshm Island in the Persian Gulf. It is a mountain made almost entirely of salt, a dazzling white spire against the desert landscape. Within its core lies one of the world's longest salt caves, a labyrinthine passage carved by underground streams. The journey through it is a descent into a world of crystalline pillars, stalactites, and glimmering walls, each reflecting the light like a thousand diamonds. It is a reminder that even in the most desolate places, the Earth holds unimaginable beauty.

The presence of these salt domes has not gone unnoticed by humanity. For millennia, salt has been a precious commodity, a source of wealth and power. It was traded along ancient routes and guarded with ferocity. But the salt domes also hold a darker, more volatile secret. They are a trap for hydrocarbons, the oil and gas that have shaped the modern world. As the salt diapirs rise, they create impermeable caps, sealing pockets of oil and gas beneath them. This geological happenstance has made the Persian Gulf region a powerhouse of energy production, but also a flashpoint for conflict.

And so, the story comes full circle. Salt holds the history of a planet, its minerals born of seas that evaporated long before the coming of man. The ancient sea, in its final, silent exhalation, left a legacy of both beauty and struggle. The salt, once a symbol of purity and preservation, now stands as a testament to the complex interplay of geology, human history, and resource extraction. It is a reminder that the seemingly lifeless Earth is in fact a dynamic, living entity, and that even the most mundane substance can hold within it the power to shape continents, economies, and the fate of nations.

8

A Salt Time Capsule

How salt's crystalline ledger of time blasts from the past

The Chehrabad Salt Mine in Zanjan, Iran, is no mere quarry; it is a tomb of immense and sudden tragedy, now a site of unprecedented archaeological drama. Deep within its crystalline veins lay a secret, a series of astonishing discoveries that would rewrite pages of ancient history. Here, the earth did not consume its dead but instead froze them in time, yielding the perfectly preserved remains of several men–relics now known to the world simply as the 'Salt Men'.

This stunning preservation is the mine's unique and terrifying gift. The initial, fateful discovery unfolded in 1993, when ordinary miners, cutting into the rock salt, suddenly unearthed a body. It was the first of these casualties, mummified not by human effort, but by the very forces of nature and geology. Since that chilling moment, five more Salt Men have emerged from the salt-laced darkness.

The true marvel, however, lies in the terrifying efficacy of their tomb. Immersed in a concentration of salt so high it defied decay, the bodies were so spectacularly preserved that even after 1,700 years, some of their tissue remained pliable and soft. This is not the dry, brittle evidence of a skeleton; this is a vivid, unblemished snapshot of ancient life, allowing scientists a terrifyingly intimate glimpse into a world otherwise lost to the dust of millennia.

The 'Prince' and a World of Clues

The first to emerge from the crystalline embrace of the Chehrabad Salt Mine was a figure so perfectly preserved that it defied the millennia: Salt Man No. 1. This was no mere skeleton; it was a ghost of history, a silent testament to a life lived and abruptly ended.

Scientific investigation, a careful tracing of time via carbon-14 dating, pulled back the curtain on his era, placing him at the crossroads of empires–the tumultuous late Parthian and early Sasanian periods, roughly 1,700 years ago. His demise had sealed him in an airless tomb of salt, protecting not just his remains, but also the vivid details of his final moments. Adorning him were the markers of privilege: gold earrings and exquisitely crafted leather boots, a fashion frozen in time. Beside him lay a trove of his world: sharp knives, a mysterious silver object, and a gleaming, precious sapphire. These weren't the possessions of a common laborer; they were the stunning proof that a man of extraordinary status—perhaps a prince or a high-ranking official—had met his dramatic end deep within the tunnels.

Yet, this magnificent discovery set the stage for a profound and agonizing tragedy. The sheer isolation of Salt Man No. 1 initially led archaeologists to a fatal assumption: that he was the mine's only secret. Based on this premature conclusion, the vital archaeological excavation was called to a halt, and the relentless, destructive rhythm of mining operations was allowed to resume. It was a catastrophic miscalculation. As mechanical extraction tore through the ancient earth, it utterly destroyed priceless historic tunnels and, with them, the potential remains of other untold victims. The silence was broken only later, by the subsequent, devastating revelation that the mine held many more souls; the bodies of five additional 'Salt Men' were eventually unearthed, a haunting confirmation of what had been lost forever. The perfectly preserved bodies and belongings of the 'Salt Men' of Zanjan remain a staggeringly valuable, yet painfully diminished, window into a world thousands of years past, their existence a dramatic reminder of the knowledge that slipped through our grasp.

The Salt Speaks

The silence of the salt seemed to hold its secrets tightly until 2004, when the roar of the miners' machinery violently tore through the initial, quiet assumptions. A new, terrifying piece of evidence was ripped from the earth: Salt Man No. 2. This wasn't merely a fragment or a piece of a puzzle; it was a complete skeleton, its tissues held in a macabre, perfect embrace by the very salt that had entombed him. Strewn around him were the grim relics of a working life–tangled ropes, battered tools, and simple pottery–painting a far more dramatic and unsettling picture than the first discovery. This man's remains didn't whisper of aristocratic ritual; they screamed of a miner's sudden, brutal end, perhaps crushed by a catastrophic mine collapse, a swift, violent consequence of the earth's fury or a deadly, unforeseen failure of safety.

The unearthing of Salt Man No. 2, and the almost immediate revelation of Salt Man No. 3, finally shattered the scientific community's skepticism. The mine was not a tomb for one, but a mass grave–a vast, frozen moment in history. A dedicated, feverish team of archaeologists descended upon the site, trading the heavy machinery of the miners for the delicate, specialized tools of their craft. The years of 2004 to 2005 marked a dramatic, intense peak of exploration, their meticulous effort rewarded with an incredible find: Salt Man No. 4. He was the most complete of all the mummies, his face miraculously preserved with soft tissue, offering an almost unbearably lifelike, heartbreaking glimpse across millennia into the eyes of a man lost to the dark.

These subsequent, breathtaking discoveries dramatically recast the Chehrabad mine's role. It ceased to be merely the resting place of a single, privileged person and became the stage for a recurring, ancient tragedy, a deadly pit where miners and others had repeatedly met their final, suffocating end. The six 'Salt Men' discovered to date have transformed the mine from a commercial operation into an archaeological wonder, a world-renowned treasure. Each new body wrenched from the salt adds terrifying new layers of complexity, deepening the mystery of the lives they led and the gruesome fate that bound them to the glittering white prison. The Chehrabad Salt Mine continues its slow, grudging surrender of history, with the 'Salt Men' proving themselves to be far more than just unfortunate laborers, their detailed scientific analysis beginning to unveil surprising origins and final, startling fates.

The Enigma of the Tragic Salt Men

The discovery of six mummified bodies, known as the 'Salt Men', has provided an extraordinary window into ancient history, transcending a simple local find to become a subject of international scientific scrutiny.

'Salt Man No. 4' is a particularly fascinating traveler's mystery. His fine attire—a long robe, woolen trousers, and a leather cloak—combined with a knife at his waist and two small jars found beneath him, suggests he was no ordinary laborer. This sophisticated clothing hints at a higher status, perhaps that of a merchant or traveler. His death, likely caused by a mine collapse, left his purpose a mystery. The jars raise a tantalizing question: was he seeking buried treasure or intending to fill them with coins before fate intervened?

The incredible preservation of the bodies led to samples being sent to Oxford University in 2006 for groundbreaking forensic analysis, including carbon-14 dating and hair analysis. The research, led by Dr. Pollard, yielded two surprising conclusions:

Analysis of the mummies' hair proteins revealed a startling truth: the Salt Men were not native to the Zanjan region. Their hair showed two distinct protein signatures: one local to the area and another indicative of a diet rich in marine organisms. This suggests they either spent time near the sea or traveled from a coastal region, such as Mazandaran near the Caspian Sea, or potentially from Tehran or Qazvin, to acquire salt.

The Oxford tests also dramatically altered the timeline of the discoveries. Salt Man No. 2 (an average-height man, aged 30-35 at death) was confirmed to date back 1,700 years to the Sasanian era. Crucially, Salt Men No. 3, No. 4, and No. 5 belonged to a much older period: the Achaemenid era (550–330 BC). This finding confirmed that the Chehrabad Salt Mine was in use for millennia, and the victims were not from a single disaster but a series of isolated tragedies spanning centuries.

A Teen's Tragic End

Initial Oxford studies on Salt Man No. 4 provided broad context, but a detailed analysis by Iranian experts unearthed a heartbreaking, personal truth: the incredibly preserved body belonged to a 16-year-old boy. His death was no mere accident; he suffered a severe heart injury from the crushing force of a mine collapse. The very rock that took his young life also became the salt that preserved his remains for millennia.

This wasn't an isolated tragedy. Historical research and local accounts from Chehrabad villagers revealed the mine was a recurring death trap, confirming multiple collapses over the centuries, including a recent one about 150 years ago that killed seven people. Spurred by the realization of the mine's deadly history, further exploration was initiated, leading to the discovery of the sixth 'Salt Man'. Initially, only the skull was found, which was left in situ for later, careful excavation in 2008.

The ongoing study of the 'Salt Men' is a powerful, poignant reminder of the high human cost of ancient industry. Each preserved body offers a vivid, tragic narrative–a life, a livelihood, and a deadly end, all buried and perfectly conserved by the salt they were mining.

A Time Capsule's International Significance

The discovery of the 'Salt Men' is a dramatic rupture in our historical consciousness that utterly reconfigures our understanding of ancient trade routes and the grueling existence of those who toiled in the mine's depths. What began as a mere curiosity, a preserved corpse unearthed by chance, has metamorphosed into a profoundly international narrative—a terrifyingly tangible intersection of commerce, existence, and ultimate tragedy. The meticulous, ongoing forensic analysis is not simply detailing facts; it's painstakingly charting the microstructure of ancient life and death, an account gaining in both detail and dramatic intensity with every new revelation. This astonishing phenomenon demonstrates the peculiar, almost mystical power of salt, which acts not merely as a culinary preservative, but as a crystalline time capsule, freezing and safeguarding the very substance of history.

9

Cities Founded on Salt

"Even salt is better than poverty."
Akan Proverb: "Nkyene koraa
te kyen ohia")

In the stark, brutal calculus of survival, salt was not merely a seasoning for the table; it was the very scaffolding of civilization, an indispensable commodity whose control mapped the rise of empires and the fates of peoples. It is not by accident that countless prominent cities throughout history sprang up near salt sources, for this white crystal held the power of life and death: it was the critical element for preserving the food that sustained armies and survived winters, the foundational currency in ancient trade networks, and the deep, rich vein from which untold wealth and power flowed.

Why Salt was So Important

Prior to the age of cooling, salt stood as the foundational technology against decay, allowing for the preservation of flesh—be it meat or fish—and a variety of other perishables. Its necessity for food preservation was absolute, for it was the fundamental enabler of human endeavor: the sustenance that propelled long-distance trade, fueled ambitious military campaigns, and, most crucially, granted populations the sheer capacity to survive the killing frost of winter. Thus, a reliable salt supply was not merely beneficial; it was the very guarantor of a community's existence and its eventual ascent.

Salt played a major economic role because its

scarcity made it a resource of immense, tangible value. The rulers who asserted dominion over the mines, the sun-baked flats, or the arterial salt trade routes accrued an economic power that was virtually unassailable. The resultant revenue, derived from pervasive salt taxes, became the bedrock of state finance, the deep treasury from which governments funded their legions, erected vast infrastructure, and ultimately forged their prosperous metropolises. It is an echo of this primal, indispensable value that survives in our modern lexicon, giving us both the measure of a person's worth—to be 'worth one's salt'—and the ubiquitous term for compensation: 'salary', both linguistic fossils testifying to salt's forgotten role as currency.

What is prefigured in the nomenclature of the ancient world's grand urban projects, in the etymological memory of the Via Salaria—that primal artery, or 'Salt Road'—is not merely an itinerary of commerce, but a testament to the agonizing necessity of the mineral. The strategic elevation of cities along these crystalline flows, proximate to the brutal, primal source, elevated them beyond mere settlements; they became, rather, crucibles of wealth, hubs whose existence is predicated upon the taste of survival.

Some might claim that it is an exaggeration that the rise of Salzburg–the 'salt fortress', named for its ancient salt mines–or the ravenous expansion of Liverpool, was from controlling the salt trade. Nowhere in the entire history of mercantile endeavor is there as tireless and detailed a record of a city's micro-structure of accumulated capital, showing how the control of this white, essential dust constitutes the body of its fortune. Salt was a catalyst for urban development. The wealth generated from salt production and trade provided the capital needed to build and develop urban centers. It funded the construction of city walls, public buildings, and markets, and it attracted merchants, artisans, and laborers.

The development of Poland's Wieliczka Salt Mine, for example, brought immense wealth to the Polish state and contributed to the funding of the Jagiellonian University. In the shadowed depths beneath the quiet Polish soil, where the cities of Wieliczka and Bochnia stand as monuments to a mineral empire, lies an astonishing truth: a colossal, crystalline labyrinth of salt, the remnant of some of the world's oldest and most profitable mines. This subterranean wonder, cut from the earth over 700 years ago and merely a short distance from Krakow, is far more than a tourist attraction. It's a monument to hard work, unwavering faith, and ingenious artistry that survived an era when its country was behind the Iron Curtain. The answer is as surprising as it is inspiring. Poland, having endured the devastation of past wars, held a deep sense of pride in its history. This pride fueled an early, visionary decision to preserve its heritage. Long before the UNESCO World Heritage designation was a global standard, Polish authorities declared in 1928 that the old mines should become a monument. This commitment led them to be one of the first to sign the World Heritage Convention in 1972, ensuring their cherished mine would be protected for all time.

The Legend of the Ring

While the mine's history is rooted in human innovation, it's also steeped in a charming tale of a princess and a prophecy. Long ago, a Hungarian princess named Kinga was engaged to a Polish duke. Eager to give her new kingdom a priceless gift, she was at a loss. But one night, a dream revealed the solution. She was told to toss her engagement ring down a salt mine shaft in Hungary and then watch for a sign. As her party journeyed through Poland, they arrived in the village of Wieliczka. Princess Kinga spotted a bubbling spring–a salt spring. Recognizing it as the sign from her dream, she instructed her miners to start digging. The suspense mounted as they worked, hoping the dream would prove true. When the first block of salt was unearthed, the miners found something glittering inside: the princess's ring. The legend says the ring had traveled through a subterranean salt deposit from Hungary, bringing the gift of the mine with it. This miraculous discovery cemented Kinga's place in Polish history; she was later canonized as the patron saint of miners.

The exquisite Francis Joseph Ballroom, which is constructed entirely of salt in the Wieliczka Salt Mines, Galicia. Even the chandeliers are polished salt crystals. Photo Albert F. Calvert

The King's 'White Gold'

Before salt was a gift from a princess, it was the lifeblood of the Polish monarchy. Polish princes and kings quickly realized the strategic value of salt extraction and established a monopoly on the industry in the Middle Ages. The mines in Wieliczka and the nearby village of Bochnia were merged into a single enterprise, the Cracow Saltworks. This move was a financial masterstroke. At times, a third of the king's entire revenue came from salt–the 'white gold'.

This incredible wealth funded the royal court, paid clerks, and built magnificent castles and churches. It even helped establish Poland's first university in Krakow in the 14th century. The mine's prosperity wasn't just for the kings; the saltworks managers themselves became influential figures, erecting impressive townhouses that dotted the region.

An Underground World of Art and Faith

Over the centuries, the miners didn't just extract salt; they sculpted their world. Driven by their deep faith, they carved intricate chapels directly into the salt. The most spectacular of these is the Chapel of St. Kinga, a cathedral-like space located 330 feet underground. Every detail, from the grand chandeliers to the altars and New Testament-inspired carvings, is made entirely of salt. It's so breathtaking and revered that it continues to host weekly Sunday services, weddings, and concerts.

But the mine's appeal goes beyond faith. It's a place for adventure and creativity. In the largest chamber, measuring 112 feet tall, thrill-seekers have made history by holding the world's first underground balloon flight and even the first underground bungee jump.

The Wieliczka Salt Mine is a testament to the Polish people's resilience and their commitment to preserving their rich culture. It's a place where history, legend, and human ingenuity come together in a unique, spectacular way, leaving visitors to wonder just how deep the layers of this fascinating story go.

Salt Cities

A number of cities, including Nantwich, Northwich, and Middlewich, have names that end in '-wich', an Old English suffix often associated with salt-making towns. Similarly, Ston, Croatia, on the Pelješac peninsula, was a planned settlement, built by the Republic of Ragusa (Dubrovnik) in the 14th century specifically to protect and profit from its valuable salt pans. Munich, Germany's founding is directly linked to the salt trade, as Henry the Lion, the Duke of Bavaria, built a bridge over the Isar river in 1158 to divert salt trade revenues away from the bishops of Freising.

Salt was just as influential in the founding and development of cities in the United States as it was in Europe and other parts of the world. The control of salt resources, production, and trade routes was a major factor in the growth of several American cities. One famous U.S. city that owes its existence to salt is Syracuse, New York. Syracuse is perhaps the most famous example of a U.S. city founded on salt. The 'Salt City', Syracuse's nickname, is a direct reference to its history. The region, particularly around Onondaga Lake, was home to a large deposit of brine springs. Native American groups, particularly the Onondaga Nation, were aware of these salt springs and used them for their survival.

Salt became Syracuse's economic engine beginning in the late 1700s, when commercial salt production took off in the area. The state of New York, which owned the salt lands, leased them to manufacturers, and the salt industry became the economic engine of the city. Syracuse's strategic location and its connection to the Erie Canal, which was even called 'the ditch that salt built', allowed it to become the country's leading salt producer in the 19th century. This wealth funded the city's growth and development.

Another city built on salt is Detroit, Michigan. Detroit has a massive and historically significant salt mine directly beneath the city. A subterranean resource, the Detroit Salt Mine is a massive underground complex, covering over 1,500 acres. Rock salt deposits were discovered in the late 1800s, and commercial mining began in the early 20th century. It still holds modern significance, since the mine produces road salt used for de-icing during winters colder than in Scandinavian cities. The presence of this immense, valuable resource contributed to Detroit's economic identity and industrial power historically as a major part of the city's industrial base.

Hutchinson, Kansas, known as 'Salt City', is another prime example of a city built on a subterranean salt deposit. Salt was accidentally discovered in Hutchinson in 1887 by a land speculator who was drilling for oil. This discovery of vast salt beds was a windfall for the city.

The salt mines developed into a solid economic foundation, as salt quickly became a major source of jobs and income, solidifying the town's prosperity. Today, the salt industry remains a key part of Hutchinson's economy, and the city even has a museum, Strataca, located in the mine itself.

Grand Saline, Texas abounds in natural salt flats. The Texas town's name, 'Grand Saline', is a direct French translation of 'large salt marsh', reflecting its origins. The area has a history of salt production dating back to the late 17th century. The salt dome beneath the town is one of the largest in the world. The salt industry was the driving force behind the town's growth. Grand Saline is the home of the Morton Salt company, which has had a major presence there for decades.

Saltville, Virginia is a town whose name and existence are entirely due to its salt deposits. During the American Civil War, the town was the site of a major saltworks that supplied a significant portion of the Confederate army's salt needs for food preservation. Because of its strategic importance, it was the site of two major battles during the Civil War, as Union forces attempted to destroy the saltworks and cripple the Confederate supply chain. The history of the town is inextricably linked to this vital resource and the conflicts fought over it.

And Salt Lake City is a unique and fascinating case when it comes to the relationship between a city and a salt source. While many cities were founded because of salt as an economic commodity, Salt Lake City's relationship with the Great Salt Lake is more about geography, isolation, and a name-based identity, rather than a direct economic foundation on the salt trade.

The key reason for Salt Lake City's founding was not salt, but as refuge from religious persecution. The city was founded in 1847 by members of The Church of Jesus Christ of Latter-day Saints (Mormons) led by Brigham Young. They were seeking a remote and isolated place where they could practice their religion freely, away from the hostilities they had faced in the eastern United States. The remote, arid valley of the Great Salt Lake was seen as an ideal location for this purpose.

The Great Salt Lake's role as a geographical landmark became important because the Great Salt Lake was a prominent and unmissable geographical feature. The name 'Salt Lake City' (originally 'Great Salt Lake City') was a simple and logical choice, given the lake's dominance of the valley. The pioneers did not settle on the lake itself, as it is too salty. They settled miles to the east, at the base of the Wasatch Mountains, where they could access vital freshwater streams for irrigation and drinking. This source of fresh water was essential for their goal of creating a self-sufficient agricultural community.

A symbol of isolation, the saline lake was a natural barrier and a symbol of the isolation the pioneers were seeking. It was a place that other westward-bound settlers, primarily those on their way to the California Gold Rush, were generally avoiding.

While salt was not the initial reason for the city's founding, it quickly became an important economic resource after the city was established. Salt played an important Economic Role in its early industry and later development. The early pioneers experimented with producing salt by boiling the lake water. This was an essential resource for their community and a way to generate income. Over time, large-scale salt production companies, including Morton Salt, built operations around the lake, with modern works using solar evaporation ponds to harvest the salt. Today, the salt and other mineral extraction from the Great Salt Lake are significant industries, providing salt for de-icing roads, industrial uses, food and medicinal production.

Salt Wiches

The term 'wich' in place names often signifies a salt-producing location, derived from an Old English word for 'salt house' or 'salt pit'. According to William Smith of Cheshire county in medieval England and author of King's Vale Royal (1656 edition), "The house in which the salt is boiled is called the Wychhouse; whence may be guessed what wych signifies, and why all those towns where there are salt-springs or salt made are called by the name of wych, viz.,

Namptwych, Northwych, Middlewych, Droitwych."

Witch Wych?

The word 'wych' also refers to a tree with pliant or bending branches, like the wych-elm, used for 'water witching' (or divining), and the wood used for the rod was called a witch rod or divining rod. The name 'witch-hazel' in North America is believed to have been given because its branches resembled the wych branch's flexible bending.

However, this function did not follow the etymology to the USA, where salt centers were named otherwise. Syracuse, New York, was nicknamed 'The Salt City' because of its immense salt production from the Onondaga Salt Springs. Utah is a major source of salt from both the Great Salt Lake and inland mines, such as the one in Redmond, which mines salt from an ancient inland sea. Detroit, Michigan, sits atop part of the massive Salina Group salt deposits and is home to a large working salt mine.

Sandwiches

This brings us around to the word 'sandwich', an eponym, meaning it's named after a person. It's believed to be named after John Montagu, the 4th Earl of Sandwich (1718-1792), an English aristocrat. According to legend, he would ask for meat to be served between two slices of bread so he could eat without using a fork and continue gambling without a break. While he was the Earl of Sandwich, a title that came from the town of Sandwich, Kent, his name is the direct source of the word for the food. Since there is no evidence of a salt factory there, you would have to go all the way to Namibia to find a Sandwich connected to salt production. The Walvis Bay Saltworks in Namibia, which is located near Sandwich Harbour, does produce salt, so much that it is the largest producer of solar-evaporated sea salt in sub-Saharan Africa. The company's saltworks are in the Kuiseb river delta, and the site is accessible through tours to Sandwich Harbour.

The Walvis Bay Saltworks in Namibia is an evaporite deposit, but it's a modern, man-made one. It's an example of a contemporary solar salt operation, not a geological deposit formed over millions of years.

10

The Detroit Salt Mine

"I kissed her salty tears and murmured,
murmured I don't know what. I felt her
body straining, straining to meet mine
and I felt my own contracting and
drawing away and I knew that I had
begun the long fall down."
—James Baldwin, *Giovanni's Room*

While the Zagros Mountains hold their salt in silent, ancient formations, the story of salt in Detroit is a tale of human ingenuity and industrial might. The Detroit Salt Mine is a subterranean world, a geological mirror to the city's industrious history, lying hundreds of feet beneath the city streets. Before the arrival of settlers, the Native American communities relied on the natural springs and the local ecology, not on the deep industrial-scale mining that would emerge centuries later to harvest the vast supply of salt beneath Detroit. The Detroit Salt Mine, in active operation since 1910, is a testament to the city's reliance on and transformation of the natural world.

The Detroit salt bed is part of the Salina Group, a vast evaporite deposit formed over 400 million years ago when a prehistoric sea covered the region. Unlike the Zagros Mountains, where salt is a product of tectonic uplift, Detroit's salt lies in horizontal layers, a flat and uniform record of a calm and ancient seabed. The salt here, primarily halite—(NaCl), or table salt—is a consequence of sustained evaporation in a shallow basin, rather than the violent, tectonic squeeze of the Tethys Sea.

The connection to the Zagros Mountains lies

in the fundamental scientific principle of evaporite formation. In both regions, immense quantities of water evaporated, leaving behind a concentrated mineral legacy. However, the geological context is vastly different. The Zagros salt is a dynamic, diapiric force, flowing and shaping the landscape above it. The Detroit salt, in contrast, is a stable, layered deposit, a static resource waiting to be mined.

A City's Hidden Heart

The Detroit Salt Mine is a marvel of engineering. Miners descend over 1,200 feet, working in a vast, dark labyrinth of tunnels. The extracted salt is used for a variety of purposes, most notably for de-icing roads in winter, a crucial part of life in the Midwest. This practical application contrasts sharply with the salt of the Zagros, which, while also having human uses, is more profoundly tied to oil and gas formation and has shaped a landscape of geological wonders and international conflict.

In essence, the Zagros Mountains represent the geological suspense of salt—a force that pushes and deforms the land, with hidden implications for energy resources. The Detroit salt mine, on the other hand, represents the uplifting, practical application of that same natural resource—a story of a city's reliance on the earth's bounty to keep its lifeblood, its roads, flowing even in the harshest winter.

Both are stories of salt, but they are told in completely different voices: one a whisper from deep time, the other a roar of human industry.

The Salina Group is a geological formation that is directly related to the Great Lakes region because it underlies most of it, specifically the Michigan and Appalachian Basins. This vast deposit of evaporite minerals, primarily rock salt (halite), was formed over 400 million years ago, long before the Great Lakes themselves came into existence.

Formation and Relationship

The Great Lakes as we know them today are a geologically recent phenomenon, formed by the scouring action of continental glaciers during the last Ice Age, which ended just over 10,000 years ago. The Great Lakes are a product of glaciation. Their vast basins were carved out by the immense pressure and scouring action of continental ice sheets during the last Ice Age.

When the glaciers retreated, they left behind massive depressions that filled with their meltwater, forming the freshwater lakes we see today. The lakes are kept fresh by a continuous flow of water that exits into the St. Lawrence River, flushing out dissolved minerals. These glaciers carved out the basins that now hold the lakes.

The Great Lakes were never saltwater seas. Their basins were carved out by glaciers during the last Ice Age, and they filled with meltwater as the glaciers retreated. The water that filled them was freshwater from the melting ice sheets and from precipitation runoff.

The reason the Great Lakes are freshwater and not saltwater is due to a simple but critical geological feature: they have an outlet. The water in the Great Lakes constantly flows out through the St. Lawrence River to the Atlantic Ocean, carrying dissolved minerals and salts with it.

A Tale of Two Systems

Lakes with outlets: These are freshwater lakes, like the Great Lakes. Water flows in from rivers, streams, and precipitation, and also flows out. This continuous flow prevents salt and other minerals from accumulating. Any salt that enters the lake from sources like ancient bedrock (the Salina Group) or road runoff is flushed out. This is a fundamental concept of hydrology. These are saltwater lakes, like the Great Salt Lake or the Dead Sea. Water flows in, but there is no river or stream for it to flow out. The only way for water to leave is through evaporation. When water evaporates, it leaves behind any dissolved minerals and salts, which accumulate over time, making the lake increasingly saline.

The existence of the Great Lakes' outlet to the ocean is what distinguishes them from saltwater lakes and is the primary reason they've remained freshwater bodies despite their immense size and the salt deposits that lie beneath them.

Other fresh water sources have recently been found off the coast of Maine. 'Offshore Aquifers' are vast freshwater aquifers beneath the ocean floor, a more mysterious and recent discovery. As the text states, scientists are still trying to determine their exact origin.

The leading theories, that either rainwater seeped into coastal sediment that was later submerged or that melting ice sheets left water trapped in the soil, both contrast sharply with the Great Lakes' formation. These aquifers are essentially 'fossil' water, sealed off from the surrounding saltwater environment, while the Great Lakes are part of an active, flowing surface water system.

Beneath the salty oceans lies a surprising source of fresh water. A team of researchers recently drilled off the coast of Massachusetts and discovered vast freshwater aquifers under the seabed. This discovery challenges the old maritime saying, "water, water everywhere, nor any drop to drink," and offers a potential new solution to the growing global demand for drinking water.

The Unexpected Discovery

The researchers found that the fresh water starts at a shallower depth and extends deeper than they had originally anticipated. While this isn't the first time fresh water has been found under the sea, the precise mechanisms of its formation remain a mystery.

One theory suggests that rainwater seeped into a beach that was later submerged by the ocean. Another posits that a melting ice sheet sank into the soil, leaving its water behind. Scientists are now testing these hypotheses with thousands of samples extracted from a massive aquifer believed to stretch from Maine to New Jersey.

Potential and Risks

The discovery of these extensive underwater reservoirs holds great promise for countries facing water scarcity. However, exploiting this resource comes with significant risks. Pumping the offshore fresh water could deplete the terrestrial supply on nearby islands, raising questions of ownership and sustainability. The history of human exploitation of underground resources is not promising, and researchers will spend months analyzing the samples to fully understand the potential consequences before any further action is taken.

The Salina Group underneath the Great Lakes, however, was created during the Silurian Period when a shallow, ancient sea covered the North American continent. Due to a warm climate and restricted circulation, this sea evaporated repeatedly, causing dissolved minerals to precipitate and form thick layers of salt, gypsum, and anhydrite.

This process is a classic example of evaporite deposition. The Salina Group is not a feature of the Great Lakes themselves but rather a product of the deep geological history of the bedrock beneath them. The salt layers are a permanent part of the foundation on which the Great Lakes region is built.

The Salina Group is a vast, Late Silurian-age

geological formation of sedimentary rock found in the northeastern and Midwestern parts of North America. It's a significant evaporite deposit, meaning it was formed from the evaporation of a shallow, ancient sea.

Formation and Composition

The Salina Group was deposited over 400 million years ago when a prehistoric sea, in an area roughly corresponding to the modern-day Great Lakes region, was cut off from the open ocean. The warm climate caused the water to evaporate, leaving behind a thick sequence of minerals. The Salina Group is primarily composed of halite (rock salt), but it also contains beds of dolomite, shale, anhydrite, and gypsum. These layers are stacked, with salt beds often alternating with the other rock types.

Economic Importance

The immense size and depth of the Salina Group make it an important natural resource.

The salt is mined from deep underground for various uses, most notably for de-icing roads in cold climates, a vital function in the Great Lakes region. The formation's impermeable salt layers also have economic value in the oil and gas industry, as they can act as a seal for hydrocarbon reservoirs.

The relationship between the Salina Group and the Great Lakes is primarily economic and practical. The Great Lakes are a major freshwater resource and a hub for transportation and industry, while the Salina Group is an immense source of salt. This resource is mined from deep underground, as seen in the Detroit Salt Mine, to be used for a variety of purposes, including de-icing roads in winter. This direct human use of the ancient Salina Group deposits makes the two seemingly disparate natural phenomena–the geologically ancient salt and the geologically young lakes–intricately linked in the modern-day life of the region.

In Detroit's industrial heart, a different kind of history is etched into the earth. While the city's story is typically told in terms of steel, automobiles, and bustling factories, a more secretive narrative unfolds 1,200 feet below the surface. This is the story of the Detroit Salt Mine, a place whose inner workings were, for generations, shrouded in mystery, echoing a nationwide trend of secrecy within the salt industry.

The salt trade, one of the oldest and most vital in the United States, has historically been an intensely private affair. In Detroit, this was especially true. The men who built and operated the mine were a small, tightly-knit group who fiercely guarded their knowledge. Their methods, expenses, and even the sheer scale of their operation were considered proprietary secrets. Outsiders were consistently rebuffed in their attempts to learn more. This culture of secrecy wasn't about malice; it was about protecting a hard-won monopoly and ensuring a competitive edge in a lucrative, if underexplored, industry.

Unearthing the Secrets

For a long time, the public knew little about the colossal underground world beneath their feet. The only glimpses were of the salt-laden trucks rumbling through the city streets. But beneath the surface, a vast labyrinth of tunnels and chambers, spanning over 1,500 acres, was being carved out of the ancient Salina Group bedrock. The silence surrounding this operation was eventually broken by diligent inquiry and a stroke of luck.

To discover the Detroit Salt Mine, one had to go beyond the sparse public records. Researchers had to gain access to private records, scientific papers, and company documents that had been previously off-limits. They had to supplement this with firsthand accounts from the miners and engineers themselves, men who had spent their lives in the cold, dark passages, and who finally began to share their stories.

The unearthing of these secrets revealed not just the technical details of a massive mining operation, but the human story behind it. It showed how a small group of determined individuals harnessed a natural resource formed hundreds of millions of years ago to fuel a modern city's growth, keeping its roads safe in winter and its chemical plants productive year-round. The opening up of the Detroit Salt Mine's story is a testament to the fact that even in an age of information, some of the most compelling narratives are still found in places where secrets were once buried deep. Similar historical examples of these strategies exist in other salt mines around the world, though often with their own unique twists.

Modern vs. Ancient Evaporite Deposits

While ancient evaporite deposits like the Salina Group in North America or the salt domes in the Zagros Mountains were formed by the natural evaporation of prehistoric seas over millions of years, the Walvis Bay Saltworks uses the same fundamental process but in a controlled, industrial setting.

The company pumps seawater from the Atlantic Ocean into a series of shallow, interconnected ponds. The hot, dry climate of the Namib Desert provides the solar energy and wind needed to evaporate the water, leaving the salt behind to be harvested. This process is highly efficient and capitalizes on the region's natural conditions to create a modern evaporite deposit. A saltwater solution can conduct electricity because salt, when dissolved in water, breaks apart into charged particles called ions.

The Role of Ions in Electrical Conductivity

In its solid form, table salt (sodium chloride, NaCl) is an ionic compound composed of a crystal lattice of positively charged sodium ions (Na^+) and negatively charged chloride ions (Cl^-). These ions are held in a rigid structure and cannot move freely, so solid salt does not conduct electricity.

However, when salt is dissolved in a polar solvent like water, the water molecules pull the ions apart, freeing them from the crystal lattice. Once released, these ions are mobile and can move through the solution. When an electrical potential is applied across the saltwater solution (e.g., from a battery with two electrodes), the mobile ions become the charge carriers. The negatively charged chloride ions are attracted to the positive electrode, while the positively charged sodium ions move toward the negative electrode. This directed movement of ions constitutes an electric current, allowing the solution to conduct electricity. This process is what makes saltwater an electrolyte, a solution containing ions that can conduct electricity. The more salt you add to the water, the more ions are present, and the more conductive the solution becomes.

11

The Uses of Salt

"Governing a large state is
like cooking a small fish."
—Laozi (often modernized
to caution against excessive effort,
suggesting one should neither overcook
the fish nor add too much salt)

Before the age of electricity and laboratories, salt was not merely a seasoning, but a vital force woven into the fabric of existence. It held dominion over preservation, medicine, ritual, and trade, and stood as an indispensable treasure whose incredibly wide-ranging uses dictated the rhythm of daily life.

In modern times, when machines do so much for us, our relationship to salt has changed, but the marriage remains solid. One of the closest-to-home uses for salt is as a metabolism booster.

The Link Between Salt and Weight

The body is naturally equipped with the tools needed to manage weight and energy. There's a simpler, safer way to burn fat and support your metabolism without the unwanted side effects of prescription drugs, according to Dr. Ben Bikman. He emphasizes that salt is an essential, fat-burning mineral, not an optional one, and that restricting salt intake can lead to an increase in insulin levels, a hormone that promotes fat storage. This can cause insulin resistance, which is a major contributor to metabolic syndrome, type 2 diabetes, and heart disease. He says that salt restriction may help with hypertension on one level, but the belief that it is healthy in all cases is ironic because it can increase insulin resistance, the very problem it's often meant to prevent. His approach begins with a mineral that has been unfairly demonized for decades: salt. Salt earned a terrible reputation due to studies that implicated its consumption as a cause of high blood pressure. As a momentary physiology lesson, it's true that if you consume a high amount of salt, your body will temporarily retain water to balance the sodium level, which could be reflected by an elevated blood pressure.

However, multiple large-scale studies have since found that telling a population with hypertension to drastically cut salt intake results in a negligible, irrelevant drop in blood pressure, often only one or two points. This is because salt is not the primary driver of high blood pressure; the true culprit is insulin resistance. Insulin resistance forces the body to retain salt and constricts blood vessels, both of which lead to high blood pressure. Bickman says that instead of advising people to cut back on salt, the true focus should be on cutting back on what spikes insulin, namely refined starches and sugars.

He urges patients to stop fearing salt, as it's not the reason you're holding onto stubborn fat. In fact, not getting enough salt may be why your body is overproducing fat-storing hormones. By adding back the right kind of salt–such as Himalayan, Celtic, or iodized sea salt–you can help your body finally stop overproducing these hormones. A simple way to start is by trying a morning glass of warm water with a pinch of natural salt, or by simply seasoning your meals with real salt. He says this is a simple, safe method that can kickstart your metabolism without negative side effects: your fat cells aren't trying to sabotage you; they are just responding to your body's needs. He says giving the body the nutrients it's been missing will activate fat-burning potential.

A vintage pamphlet described the "One Hundred and One Uses for Salt," as per the Diamond Crystal Salt Company, based in St. Clair, Michigan. These booklets were published in the early 20th century, with editions dating back to at least 1914 and 1922.

The pamphlet is more than just a list of recipes; it highlights the incredibly wide-ranging uses for salt in daily life before modern chemistry and home appliances became commonplace. While it's difficult to find a complete and verified list of all 101 uses, sources that have described or reproduced parts of the pamphlet reveal some of the fascinating and practical tips included.

Some examples of what the pamphlet covered, with the uses are categorized:

For the Kitchen and Cooking

Brining for pickles: The pamphlet advised making a brine strong enough to float an egg.

Preventing meat spoilage: Salt was a primary preservative for meat, a common practice before refrigeration.

Keeping butter and hard cheese dry: Salt absorbs moisture, which was useful for preservation.

Preserving color in boiled vegetables: Adding salt to the water helps vegetables retain their green color.

Preventing odors: Putting salt in the water when boiling cabbage or other strong-smelling vegetables was recommended to reduce the odor.

Making pancakes without smoke or odor: The pamphlet claimed this was possible with salt.

Around the House

Cleaning brass and copper: Salt was used as a cleaning agent for metalware.

Freshening withered apples: Soaking them in a salt solution was a tip to restore their crispness.

Making brooms and toothbrushes last longer: Soaking these items in saltwater was suggested as a way to extend their lifespan.

Cleaning a greasy stove: Salt was recommended as a way to clean a stove top.

Extinguishing a fire: Salt could be used to put out a grease fire.

For the Laundry

Preventing clothes from freezing on the line in winter.

Setting colors in fabrics.

Removing stains, such as fruit stains, from clothing.

Miscellaneous

Making cut flowers last longer.

Keeping eggs fresh by packing them in salt.

These are just some of the many examples. The pamphlet served as a marketing tool to demonstrate the value of their salt beyond simple seasoning, emphasizing its versatility and importance in a pre-industrial-era home. It highlights how integral salt was for a range of uses, from preservation, to cleaning, and everyday tasks. The many aspects of salt have been felt on an intuitive level ubiquitously throughout humankind. In the words of August Strindberg from his seminal, *A Dream Play* (1902):

> I once asked a little boy why the sea was salt, and the boy, whose father was away on a long journey, said right away, "The sea is salt because the sailors cry so much."
>
> "But why do the sailors cry so much?" I asked.
>
> "Because," he said, "they always have to go away from home–and that's why they're always drying their handkerchiefs up on the masthead!"
>
> And then I asked him, "But why do people cry when they're sad?"
>
> And he said, "That's because they

have to wash the glasses of their eyes so they can see better."

Though salts, in the lexicon of modern invention, have been seemingly outmoded by newer, dazzling compounds, yet their definition lies in an unexpected depth of influence–a subtle, systemic establishment. To understand salt is to recognize that the simplest compounds often conceal the most profound, fundamental power, making its ubiquitous presence a masterpiece of both chemistry and survival.

Can salt be used as fuel?

Generally, salt itself cannot be used as a direct fuel for combustion like wood or gasoline, because it has already released its chemical energy. Instead, salt is used in a few key ways to produce, conduct, store, or carry energy, mainly in batteries.

Salt is used in Hydrogen Production, and Fuel Cells. Researchers are exploring methods to use saltwater to produce hydrogen fuel through a process called electrolysis. While traditional electrolysis requires purified water, new technologies are being developed that can use saltwater directly, making hydrogen production from a widely available resource more feasible.

Salt is an even more essential component in saltwater fuel cells, which generate electricity through a chemical reaction between a magnesium electrode and oxygen in the air, with the salt acting as a catalyst to facilitate the flow of ions, detailed in Chapter 22.

Salt for Energy Storage

Salt is now being used on a large scale to store energy, particularly in concentrated solar power (CSP) plants. A field of mirrors focuses sunlight to heat large tanks of molten salt to extremely high temperatures. This hot salt can then be stored in insulated tanks for hours or even days. When electricity is needed, the stored heat from the molten salt is used to produce steam, which drives a turbine to generate electricity, even after the sun has set.

Salt batteries, more formally known as sodium-ion batteries, are a type of rechargeable battery that uses sodium ions (Na^+) as the charge carriers. They're being developed as a promising alternative to lithium-ion batteries, primarily due to the abundance and lower cost of sodium compared to lithium. Sodium-ion batteries are already available at various suppliers, particularly those based in China. These are often marketed to DIY enthusiasts, researchers, and small-scale integrators.

Sodium-ion batteries function on a similar principle to their lithium-ion counterparts. During charging, sodium ions move from the cathode to the anode, and during discharge, they move back, generating a flow of electrons that provides power. The key difference is the use of sodium instead of lithium. While sodium-ion batteries may not completely replace lithium-ion batteries in all sectors, their safety, low cost, and material abundance make them a strong contender for large-scale energy storage solutions like supporting renewable energy grids, where weight and size are less of a concern.

Salt as a Fuel Carrier

In advanced nuclear reactor designs, like Molten Salt Reactors (MSRs), salts are used as a fuel carrier. Nuclear fuel, such as uranium or thorium, is dissolved directly into a molten salt. This liquid mixture then serves as both the fuel and the coolant for the reactor. This design offers safety benefits, as the liquid fuel can be drained away in an emergency, stopping the nuclear reaction.

Fueling the Human Body

Salt is considered an 'elixir of life' due to its

crucial role in maintaining essential bodily functions, its historical importance in human civilization, and its multifaceted applications. It is not a magical cure-all, but its fundamental necessity for survival and health makes it a vital substance.

Salt, primarily in the form of sodium chloride (NaCl), is a fundamental component of the human body. It is essential for:

- Fluid Balance and Hydration: Sodium and chloride ions are key electrolytes that regulate the amount of water inside and outside our cells. Without proper salt levels, the body cannot maintain this balance, leading to dehydration or fluid retention, both of which can be life-threatening.
- Nerve Function: Sodium ions play a critical role in the transmission of nerve impulses and muscle contractions. They are responsible for generating electrical signals that allow your nerves to communicate and your muscles, including the heart, to function correctly.

- Nutrient Absorption: Salt is necessary for the proper absorption of nutrients in the small intestine. For example, glucose absorption is a sodium-dependent process, highlighting salt's role in energy uptake from food.
- Stomach Acid Production: Chloride ions are a key component of hydrochloric acid (HCl), a vital substance in the stomach that aids in digestion and helps kill harmful bacteria.

Maintaining Electrolyte Balance

Sodium and chloride are essential electrolytes for proper nerve and muscle function, and for regulating the body's fluid balance. When you are sick, injured, or unable to eat or drink, your body's electrolyte levels can become dangerously low or imbalanced. An IV of normal saline helps to maintain these vital electrolyte levels, ensuring that your cells and organs, especially the brain and heart, can function correctly.

Salt is considered an 'elixir of life'. Not only

has it been a cornerstone of societal development throughout history, it is essential for human survival. The ions from salt–primarily sodium (Na+) and chloride (Cl−) ions, along with potassium (K+) ions–are essential for creating and transmitting the electrical signals that the brain and nervous system use to communicate. This process is known as an action potential.

The Resting Potential

Neurons, or nerve cells, maintain a difference in electrical charge across their membranes. This is called the resting membrane potential. The cell membrane has special proteins called sodium-potassium pumps that work continuously to pump sodium ions out of the cell and potassium ions into the cell. This creates a higher concentration of positive sodium ions outside the cell and a lower concentration of positive potassium ions inside, making the inside of the cell relatively negative compared to the outside. This imbalance of ions creates a stored electrical potential, similar to a battery.

When a neuron receives a strong enough signal from another neuron, it reaches a threshold that triggers an action potential. This is a rapid, temporary change in the membrane's electrical potential.

Voltage-gated sodium channels in the cell membrane open, and due to the strong concentration gradient, a flood of positively charged sodium ions rushes into the cell. This influx of positive ions causes depolarization: the inside of the cell becomes positively charged, creating an electrical impulse. As the cell becomes positive, the sodium channels close, and potassium channels open. Positively charged potassium ions rush out of the cell, resulting in repolarization, causing the inside of the cell to quickly return to a negative state.

During a refractory period, the neuron briefly

becomes even more negative than its resting state before the sodium-potassium pumps restore the initial ion balance, preparing the neuron to fire again. This wave of depolarization and repolarization, powered by the movement of ions, travels down the length of the neuron's axon to transmit information.

Rehydration and Fluid Replacement

When you're admitted to a hospital, one of the first things a healthcare provider will do is start an intravenous (IV) line and connect you to a bag of fluid. The most common fluid used is normal saline, which is a 0.9% solution of sodium chloride (NaCl) in sterile water.

The reason for this is that normal saline is an 'isotonic' solution, meaning it has the same concentration of salt and other solutes as your blood. This makes it the most effective and safest fluid for several critical medical purposes: Many medical conditions, from severe dehydration due to vomiting or diarrhea to fluid loss from a major injury, can lead to a dangerously low volume of blood and body fluids (hypovolemia). Normal saline is used to quickly and efficiently replenish this lost fluid. Because its salt concentration matches your blood, it helps to restore the fluid volume in your bloodstream without causing a dangerous shift of fluids into or out of your cells.

A Vehicle for Medications

An IV line provides a direct and immediate route for administering medications, which is often crucial in a hospital setting. Normal saline is the most common 'carrier' fluid used to dilute and deliver a wide range of intravenous drugs, from antibiotics to pain medications. It is a stable, sterile, and safe solution that doesn't interfere with the medications, whereas giving pure water intravenously would be incredibly dangerous.

Pure water has a much lower concentration of solutes than blood. This would cause your red blood cells to rapidly absorb the water through osmosis until they swell and burst (hemolysis), which is a life-threatening event. Normal saline prevents this by ensuring the fluid entering your bloodstream is in balance with your body's natural fluids.

In short, normal saline is the default IV fluid because it's a safe and effective way to quickly rehydrate a patient, maintain a stable internal environment, and provide a means for delivering essential medications.

Salt and Dehydration

Dehydration is a state where the body loses more fluid than it takes in, disrupting its normal functions. When you are dehydrated, the use of a normal saline IV solution is directly related to restoring your body's vital balance.

Dehydration is a loss of more than just water. While we often think of dehydration as simply a lack of water, it's more complex than that. Your body's fluids–including blood, sweat, and digestive juices–contain essential electrolytes, with sodium being the most important. Dehydration from conditions like severe vomiting, diarrhea, or excessive sweating results in the loss of both water and these vital salts.

This is where a normal saline IV comes in. A normal saline solution is a 0.9% solution of sodium chloride (NaCl) in water. Its effectiveness lies in its isotonic nature, meaning it has the same salt concentration as your blood. This is crucial for several reasons:

When a person is severely dehydrated, their blood volume decreases, which can lead to low blood pressure, dizziness, and a rapid heart rate. The saline solution, because of its salt content, is designed to stay within the blood vessels (the intravascular compartment) rather than immediately diffusing into cells resulting in immediate volume restoration. This rapid expansion of blood volume helps restore blood pressure and improve circulation to vital organs.

Electrolyte Balance

Dehydration often leads to an imbalance of electrolytes. The sodium and chloride ions in the saline solution directly replenish the electrolytes that have been lost, helping to restore the balance needed for proper nerve, muscle, and brain function.

If a person were given pure water intravenously, the water would rush into the red blood cells, causing them to swell and burst (hemolysis). The isotonic nature of saline prevents cell damage: by ensuring the fluid is in equilibrium with the blood, preventing harmful fluid shifts and protecting the integrity of your cells.

In essence, a normal saline IV is a precise and rapid way to address the two main problems of severe dehydration: lack of fluid volume and depleted electrolytes, ensuring the body can return to a stable state.

12

The Symbolism of Salt

"Where one has eaten salt, one can be honest." (Där man har ätit salt, där kan man vara ärlig.)
—Anon, a Swedish saying that alludes to the old practice of sharing salt (a meal) as being bound by loyalty and respect.

A crystalline whisper, a taste of paradox. Salt. It is the sting of a fresh wound and the flavor of a feast. It is the grain that makes life possible, and the essence that destroys what it preserves.

In its raw form, salt is a silent, unyielding

stone, a mineral of the earth. But once dissolved, it becomes a conductor, a catalyst. It's the very force that fires the brain's thoughts and commands the heart's rhythm. It is the language of life's electrical impulses, a bridge between chemical and consciousness.

Salt is the memory of the sea. It is the tears of grief and the tears of joy. It is the residue on the skin after a long day of labor, a testament to exertion. It is the silent signature of the ocean on every shore, the endless cycle of evaporation and return. The human body holds a salinity that echoes the primordial ocean from which we crawled, a constant, salty reminder of our origin.

In lore, salt is a ward against evil. It is the barrier a witch cannot cross, the pure substance that repels corruption. It is cast over a shoulder to blind the devil, a small act of superstition born from a deep-seated belief in its power to purify. Yet, it is also a symbol of destruction—

the biblical wrath that turned Lot's wife into a pillar of salt, an eternal monument to a forbidden glance back at the past.

Milton uses poetic references to the sea and salt in *Paradise Lost,* using the sea as a metaphor for chaos, an untamed and dangerous place that exists in contrast to the ordered and perfect world of Heaven and Earth. The 'salt' of the sea is part of its wild and destructive nature. It is a symbol of bitterness, purification through suffering, and the corrosive power of fallen nature. This is in contrast to the divine, fresh waters of Eden. In Book 7 of his epic poem, when Milton describes the creation of the world, he talks about God setting boundaries for the sea, keeping its salty flood in check. The salt and chaos of the ocean are contained so that the peaceful, fruitful land can exist.

> ...till inundation rise
> Above the highest hills: Then shall this mount
> Of Paradise by might of waves be moved
> Out of his place, pushed by the horned flood,

With all his verdure spoiled, and trees adrift,
Down the great river to the opening gulf,
And there take root an island salt and bare,
The haunt of seals, and orcs, and sea-mews' clang:
To teach thee that God attributes to place
No sanctity, if none be thither brought
By men who there frequent, or therein dwell.

– John Milton, *Paradise Lost*

This reflects a larger theme in the poem: the triumph of order and divine will over primordial chaos and rebellion. It is the foundation of ancient pacts. A 'covenant of salt' was unbreakable, a promise sealed with a substance that resists decay, a symbol of permanence and loyalty. To share salt with a stranger was the ultimate sign of hospitality, a bond forged at the table that made them kin. To be 'worth your salt' speaks to the value of labor, a direct link to the days when Roman soldiers were paid in this precious commodity.

Ultimately, the symbolism of salt is a duality. It is the essence of life and the specter of death. It is preservation and corrosion. It is a simple crystal, yet it holds within its molecular structure the power to connect, to destroy, to purify, and to endure. It is the taste of the world itself.

Religious rituals using salt & their meanings

Salt is a substance rich with religious symbolism, primarily associated with themes of purity, preservation, and protection. Across various faiths and cultures, its simple crystalline form is used in rituals to cleanse, bless, and ward off evil. Some examples of religious rituals using salt and their meanings:

The Bible (Leviticus 2:13) refers to a 'covenant of salt', signifying a lasting and incorruptible bond. The preservative nature of salt is a metaphor for the enduring and eternal nature of God's promises and His relationship with His people.

In some historical and traditional Catholic

and Anglican rites of baptism, a small taste of salt was placed on the lips of the infant or catechumen. This symbolized the 'salt of wisdom' and the 'salt of the earth' (Matthew 5:13), representing the cleansing of the person from sin and their new role as a preserver of good in the world.

The Catholic Church has a tradition of exorcising and blessing salt and then adding it to water to create holy water. The salt is believed to amplify the purifying properties of the water and to drive away evil. This mixture is used for blessings and for protection against malevolent forces.

In Japan, purification rituals (harae), salt is a primary agent of purification in Shinto. Sumo wrestlers, for instance, throw handfuls of salt into the ring before a match to purify the space. Similarly, it is common to see small conical piles of salt (morishio) placed outside the entrances of restaurants, shops, and even homes. This ritual is believed to ward off evil spirits and attract good fortune and customers. After attending a funeral, people in Japan often throw salt over their shoulders before entering their home. This is a ritual of purification (shubatsu), intended to cleanse themselves of any negative energy or lingering spirits that may have followed them from a funeral, preventing them from entering the house.

Wicca and folk traditions use salt in casting a circle. Salt is a key component in many modern pagan and Wiccan rituals. It is often used to cast a protective circle around a sacred space, creating a barrier that purifies the area and keeps out unwanted negative energies or spirits. The salt represents the element of Earth, providing grounding and stability.

Salt is widely used in folk traditions for cleansing and protection. It can be sprinkled on windowsills and doorsteps to create a barrier against malevolent spirits, or placed in bowls in the corners of a room to absorb negative energy. The act of throwing spilled salt over one's left shoulder is a popular superstition believed to blind a demon lurking behind you, stemming from the belief that evil resides on the left side.

In the Hindu Tradition in India called the Aarti Ceremony, salt is used during the worship involving lamps. Salt mixed with other substances is sometimes passed around the person or deity being honored in a ceremony and then discarded. This act is believed to absorb and remove 'evil eye' or negative energies.

Many household rituals involve salt kept in a small bowl in a corner of the home, especially at the entrance. It's believed to absorb negative energy and prevent it from entering the house. The salt is typically changed every week or so to maintain its purifying effect.

Indigenous American traditions, particularly among the Navajo (Diné) people, demonstrate a profound cultural and spiritual significance for salt, utilizing it in key ceremonies not just as a commodity, but as a symbol of generosity, life, and connection. One of the most important ritual uses is during the Baby's First Laugh Ceremony, which marks the child's readiness to fully transcend the spiritual world and join their earthly family. The person who elicits this first laugh is responsible for organizing the ceremony, where the baby, acting as the host, ceremonially distributes rock salt and gifts to all guests. This act of giving salt is a sacred lesson in generosity, meant to encourage the child to be liberal with their laughter, joy, and giving spirit throughout their life. Beyond this, salt is also used in other rites of passage, such as puberty and marriage ceremonies, and is involved in the preparation of sacred objects like buckskins for use in various chantways, highlighting its role in consecration, protection, and ensuring prosperity. This reverence for salt links the Diné to sacred salt lakes, such as the Zuni Salt Lake in New Mexico. They are sacred cultural resources associated with deities like the Salt Woman, further cementing the mineral's importance within the spiritual landscape.

The Salt Cathedral of Zipaquira′ is a fully functional Roman Catholic church built deep within the tunnels of a former salt mine, located near Bogotá, Colombia. Far more than a simple cavern, this architectural and engineering marvel is carved out of a massive halite salt rock mountain, descending over 200 meters underground. The cathedral features stunning illumination that highlights the various chambers and subterranean chapels, including fourteen small chapels representing the Stations of the Cross, which eventually lead to the massive main dome. The most unique and poignant ceremony held here is the Sunday Eucharist mass, where the combination of the vast, dark space, the shimmering salt walls, and the echoing acoustics creates a deeply spiritual and otherworldly atmosphere, attracting both local worshippers and international tourists seeking a transcendent experience beneath the Earth's surface.

14

Tissue Salts

"Man is generated of earth and air. The activity of plants called him to life. The corpse is decomposed into air and ashes, and through the vegetable world it then develops new forces in new forms." —Dr. Moleschott, Professor of Physiology in the University of Rome, *Kreislauf des Lebens (The Cycle of Life).* (1852)

The history of tissue salts is a fascinating journey that begins with ancient medical philosophy and progresses through key scientific discoveries, culminating in the development of a modern form of 'alternative medicine'.

The Philosophical Foundations: Hippocrates and the Humors

The story of tissue salts is rooted in the philosophies of Hippocrates (469-377 BC), the 'Father of Medicine'. He established the theory of humors, which posited that the human body was composed of four cardinal fluids: blood, phlegm, yellow bile, and black bile. Hippocrates believed that a person's temperament–sanguine, phlegmatic, melancholic, or choleric–was determined by the unique balance of these humors. He argued that disease was a result of an imbalance in these humors and that health could be maintained by a harmonious relationship between the body and its environment.

When Hippocrates famously advised to let food be thy medicine, he laid the philosophical groundwork for using natural substances to restore bodily balance.

The Scientific Leap: Cell Theory and Cellular Pathology

Centuries later, the understanding of the human body was revolutionized by the development of cell theory. In 1839, Matthias Jakob Schleiden and Theodor Schwann proposed that all living organisms are composed of cells and that these cells are the fundamental units of life. This groundbreaking theory established that an organism's vital functions occur within its cells and that all cells originate from pre-existing cells.

This was further advanced by the work of Rudolf Ludwig Karl Virchow (1821-1902), a German physician and anthropologist widely regarded as the founder of cellular pathology. Virchow famously stated that "every cell arises from a pre-existing cell," and he applied this principle to disease, arguing that most human illnesses could be traced to the dysfunction of individual cells. His work shifted the focus of medicine from the humors of the entire body to the microscopic processes occurring within cells, creating a new basis for understanding and combating disease.

The Birth of Tissue Salts in Germany

The stage was set for Dr. Wilhelm Heinrich Schüßler (1821-1898), a German physician who practiced homeopathy under the tutelage of its founder, Samuel Hahnemann. Schuessler was deeply influenced by the cellular pathology of Virchow. He theorized that illness was not just a cellular dysfunction but a direct result of a deficiency of certain inorganic mineral salts, which he called 'tissue salts' or 'cell salts', within the cells themselves.

When it comes to metals and other minerals

in the human body, some are more bioavailable than others, and certain ones benefit particular organs. The knowledge of these two areas was advanced in the 1870s when Schüßler set up his experiment to find out which minerals were most easily absorbed by which of the body's organs. He posited that mineral deficiencies can lead to specific health issues directly related to the mineral's function. His work aided with the understanding that a lack of the metal, calcium, can result in weaker bones since the mineral is essential for skeletal health; likewise, an iron deficiency can cause anemia, a condition characterized by fatigue and weakness, because iron is a vital component of red blood cells and hemoglobin.

Human Ashes Tell the Secret

First, Schüßler identified which tissues were made up of which salts by actually burning human organs! Next, he analyzed the ashes of each organ for their 'inorganic constituents', including their exact mineral salts. By identifying exactly which form of the essential minerals were relevant to which organs, he was able to go on to develop biochemic tissue salt therapy, also known as Schuessler Cell Salts, and published his influential work, *An Abridged Therapy,* in the mid-1870s, detailing a system that he believed was more scientifically grounded than its homeopathic predecessor. This work details his own simplified healing system, which he called 'Biochemie'.

Thus, Dr. Wilhelm Heinrich Schüßler, who was born in 1821 and practiced in Oldenburg—who had initially received his medical license in 1859 and was formally educated in homeopathy only to become increasingly disillusioned with the complexity of its remedies —is credited as the founder of the system of therapeutics known as Biochemic Medicine or simply Biochemie, from which the concept of tissue salts originated.

The central tenet of Schüßler's system is that a proper balance of twelve essential inorganic mineral salts is crucial for the normal function of organs, tissues, muscles, and cells. The theory posits that disease is not an independent entity but is caused by a deficiency or disruption in this balance, which weakens the body and makes it susceptible to illness. According to this doctrine, replenishing the deficient salt restores cellular function, eliminates symptoms, and re-establishes health and balance.

This scientific foundation is powerfully captured in an analogy offered by Schüßler himself, which compares the body's relationship to its blood to a plant's relationship with the soil. He stated that just as "...poor exhausted soil will produce only weak, sickly plants, poor blood lacking its essential constituents will produce disease-prone and weak sickly bodies." This simple, intuitive concept—that a deficiency in a foundational substance leads to illness and that its restoration brings health—has a compelling logical appeal, much like the modern understanding of nutrient-deficiency diseases such as scurvy or iron-deficiency anemia.

However, while the therapy is presented as a 'substitution therapy' for mineral compounds, a concept that aligns with Western nutritional science, it was also developed by a physician who had been trained in and practiced homeopathy. This background is reflected in the preparation of the salts, which are formulated as 'micro-doses' and prepared according to homeopathic pharmaceutical principles, often at a 6DH (Decimal Hahnemannian) homeopathic dilution (the higher the number, the more diluted; the lower, the more concentrated).

On one hand, the use of these extremely small amounts, a core principle of homeopathy known as the 'law of minimum dose', arguably stands in direct opposition to the concept of a true 'substitution therapy', which would require a significant, measurable dose of the substance to correct a deficiency. Pundits argue that this blend of a plausible-sounding rationale with an implausible delivery mechanism is a central tension in the system. On the other hand, tissue salts are in a form that is bioidentical with the form of minerals found in the human body, so they're more readily absorbed, justifying a need for less.

But more importantly, smaller doses might be better in some cases for this reason: the industry demand for scientific proof forces the medical industry to establish dosage recommendations that often place individual patients at a significant risk of overdose. These high doses are necessary to generate statistically irrefutable evidence of a drug's mechanism or efficacy in clinical trials, but they frequently exceed what a patient's body can safely or comfortably tolerate in a real-world setting, as discussed in a 2016 study in *Pain Medicine,* entitled, "Cohort Study of the Impact of High-Dose Opioid Analgesics on Overdose Mortality," by Webster, et. Al.

Research by leader of the Angiogenesis Foundation, Dr. William Li, points to the danger of taking even supplements every day, warning that the interplay between vitamin D_3 and K_2 and calcium is potentially dangerous, as they dramatically increase the amount of calcium absorbed into the bloodstream. For individuals with pre-existing kidney issues or those taking additional calcium, this elevated calcium load can overwhelm the system, potentially accelerating the damaging calcification of soft tissues and blood vessels.

Balance and common sense need to be applied. Just as we know that maintaining adequate sodium intake supports healthy nerve impulse transmission, we also know it's not a good idea to eat so much salt that we can immediately see observable scientific proof of the fact.

Disclaimer: The information provided is for educational purposes only and is not intended to diagnose, treat, cure, or prevent any disease. Always consult with a qualified healthcare professional before making any changes to your diet or health regimen.

15

Microdosing

"Let food be thy medicine
and medicine be thy food."
— Hippocrates

Hippocrates' proclamation of food equaling medicine illuminates the daily act of eating as the original microdose, a continuous titration of complex biochemistry into our bodies. Unlike the concentrated pharmaceutical capsule, which delivers a maximally short-term provably forceful medicine that can trigger a crisis in a sensitive body, a normal serving of whole food is a gradual, gentler infusion. Each bite is a delivery system for minute, natural quantities of compounds—vitamins, phytonutrients, and trace elements—that engage the body's systems softly and naturally. This constant, low-level integration ensures that the blood-brain barrier is never overwhelmed and the delicate capillary network is only ever nourished, never assaulted. Thus, food, in its natural form, operates as the ultimate, safe microdose, offering sustained health by honoring the body's need for gentle, consistent support.

The effectiveness of using smaller doses (microdosing) for better absorption and fewer side effects is best demonstrated in the brain. The scientific rationale for microdosing is strongly rooted in the documented difficulty of substances crossing the blood-brain barrier (BBB), which is formed by the unique structure of brain capillaries. The brain maintains an almost impenetrable fortress through the Blood-Brain Barrier (BBB), operating under an iron-clad policy of high selectivity. This defense system creates a severe size and property bottleneck for any foreign substance seeking entry.

The statistics alone are staggering: an estimated 98% of small-molecule drugs and virtually 100% of large-molecule drugs are refused passage by this meticulous gatekeeper. To have any chance of passive entry, a substance must be highly lipid-soluble (fat-loving) and maintain an extremely low molecular weight, generally cited as less than 400 daltons. If these physical criteria aren't restrictive enough, the brain capillaries possess sophisticated, active security measures. They are armed with drug efflux transporter proteins, notably P-glycoprotein, which act like vigilant bouncers. These transporters effectively recognize and pump substances that manage to sneak into the endothelial cells and right back out into the bloodstream, ensuring the central nervous system remains pristine.

As trained chef and American Chiropractic Neurology Board certified chiropractor in neurology Dr. Michael Pierce points out in *The Human Condition, Micro Dosing Supplements:* The practice of using smaller doses, or microdosing, especially for sensitive individuals, finds its grounding in two key scientific rationales related to this barrier. The first is a simple matter of defense: Reducing total toxic load. For a patient who is already sensitive or battling conditions like cerebral small vessel disease, their brain's defenses may be partially compromised. The entire goal of administering a smaller dose is to sharply reduce the total concentration of the substance circulating in the blood. This strategy minimizes the chance that a harmful quantity—or its dangerous, reactive metabolites—will successfully cross the weakened BBB, thus preventing a negative neurobiological response like inflammation or a systemic flare-up that could set back their recovery for weeks.

The second rationale is a clever exploitation of the BBB's own mechanisms: Avoiding saturation of efflux systems. While the term 'microdosing' is used in drug development to mean testing extremely low, sub-therapeutic amounts, it illuminates a crucial safety concept for sensitive patients. Research shows that at these microdoses, a substance may have very limited penetration precisely because the efflux systems are so efficient at clearing it. Interestingly, to achieve a therapeutic effect with certain drugs, doctors may actually need higher doses to overwhelm and saturate these efflux pumps. However, for a sensitive patient introducing a new supplement, the opposite is true. Using a microdose is a cautious tactic to ensure the substance is easily and quickly cleared from the bloodstream. By not taxing the body's detoxification or barrier systems, this approach ensures the absolute minimum risk of a major adverse event, transforming the potent defense of the BBB into a tool for initial tolerance testing.

The Burden of Proof

The individual patient, navigating the complexities of modern medicine, often encounters a profound and sometimes dangerous disconnect between the rigorous, mechanistic demands of clinical science and the simple, lived reality of their own body. This tension is never more apparent than in the realm of drug and supplement dosing, where the physician's need to prove the mechanism can override the intuitive logic of seeking the gentlest, most effective dose, leaving the patient to bear the acute burden of scientific overreach. This neglect brings new meaning to the jargon 'the burden of proof', showing that it ultimately falls not upon doctors, but upon patients.

When a researcher or doctor sets out to investigate a new compound–be it a synthetic drug or a naturally occurring vitamin like the one discussed–their primary professional obligation is not to the comfort of the individual, but to the statistical significance of the data. They are not asking, What is the smallest dose that will help this specific person feel better? but rather, What is the minimum dose required to definitively activate the targeted biological pathway in the majority of a diverse test population? To satisfy this objective, the dose must be high enough to create a clear, unmistakable signal—a 'hit'—that rises above the natural noise of the human body, individual genetic variability, and the limitations of measurement tools.

This pursuit of a definitive signal is where the dosage escalates into what a patient perceives as excessive amounts. The scientific imperative is to ensure target engagement—to flood the system with enough of the compound to guarantee that the intended biological receptor or enzyme is fully occupied. If a drug is intended to reduce inflammation via a certain cellular receptor, the researcher must administer a dose so large that the receptor is engaged regardless of whether the patient is a rapid metabolizer or a slow metabolizer. This high-water mark of dosing is a tool of the scientific method; it is a way to isolate a variable and prove a concept. However, for the individual on the receiving end, this intentional overdose–the necessary inhuman amount–is the source of severe side effects. The patient's painful experience, such as the alarming physiological reaction to a single capsule, is the collateral damage of this pursuit of proof.

The body is not a neat, standardized laboratory model; it is an integrated, highly sensitive organism. When a patient takes a dose calibrated for the statistical average or the scientific maximum, they may be receiving an amount exponentially higher than what their unique system requires. This is not simply discomfort; it is a temporary poisoning. The acute reaction–the flushing, the dizziness, the nausea–is the body's warning system going into overdrive, reacting violently to a concentration it cannot process, proving beyond doubt that the dose is indeed pharmacologically active, but at the expense of well-being and safety.

This brings us to the core issue of why doctors avoid what patients might term a microdose. A microdose, in a nutritional or intuitive sense, is the dose that provides a subtle, long-term benefit with no adverse effects. However, for the medical establishment, a microdose is largely irrelevant because it has not been validated. Medical practice is defined by the Standard of Care, which is a set of guidelines and treatments widely accepted by the healthcare community as proper. A drug's or supplement's effectiveness must be backed by high-quality, randomized controlled trials. If a microdose of a compound has never been formally studied, a physician who prescribes it is practicing outside the standard of care. They are gambling on an anecdote rather than relying on evidence, thereby incurring legal and ethical risk. The primary obstacle is not a doctor's stubbornness, but the lack of evidence for efficacy at low levels.

Research into subtle, non-toxic, long-term benefits is expensive, difficult to design, and lacks the dramatic, marketable outcomes that drive funding. A study showing a $500 dose of a compound can immediately block a key receptor is a scientific success. A study showing a $10 dose provides a slight, unmeasurable improvement in quality of life over ten years is a financial and methodological nightmare. Consequently, the only doses with established therapeutic protocols are the high, 'active' ones.

Furthermore, the higher dose serves as a crucial boundary line for safety. In the name of protecting the public, medicine must know the maximum tolerated dose (MTD). By testing at the edge of toxicity, researchers define the dangerous territory. This information, gleaned from the suffering of a few, allows doctors to proceed with extreme caution when they prescribe a high dose, mandating blood tests, liver function checks, and frequent monitoring.

Conversely, if an individual patient decides to self-administer a high-dose supplement based only on trial data, they bypass all those safety guardrails, dangerously ignoring the hard-won data on liver failure or other severe, cumulative side effects that often manifest slowly, weeks or months after the initial acute reaction has subsided.

Ultimately, the individual patient is caught in a bind: the gentle dose they desire is medically unproven and therefore inaccessible through conventional means, while the proven dose is a pharmacological sledgehammer designed to satisfy a statistical equation rather than their individual physiology.

The medical field's focus on mechanism first

ensures that when a drug is approved, it is definitively known to work, but this necessary reductionist approach often overlooks the nuanced, personalized dosing that could offer genuine relief with minimal side effects. For the patient, the key is to recognize that the dose in the trial is a scientific tool, not a clinical recommendation, and that any therapeutic use of potent compounds must be approached only under the strict, watchful eye of a healthcare provider capable of managing the established risks of inhuman amounts.

The Long Game

The profound simplicity of tissue salts therapy as a complementary healing approach rests on the idea that the secret to vitality lies in the delicate balance and proper distribution of 12 essential minerals at the cellular level. This method, often known as 'cellular biochemistry' or mineral therapy, is rooted in the Western concept that our health is fundamentally tied to these vital inorganic substances, or 'cell salts'. They are not merely trace elements; they play a critical and active role in essential bodily functions, including metabolic pathways, enzyme function, and the swift transmission of nerve impulses. The core theory suggests that any deficiency or improper arrangement of these compounds–the fundamental building blocks of the body–leads directly to cellular dysfunction and, consequently, the onset of illness.

The therapeutic applications of these salts are highly specific, offering a unique functional perspective on health. For instance, Ferrum phosphoricum is often dubbed the 'oxygen carrier', typically sought in the early stages of inflammation or fever due to its role in the body's oxygen transport. Magnesia phosphorica earns the name 'painkiller' for its traditional use in soothing spasms and neuralgic pain, while Kalium muriaticum is associated with mucosal and glandular health, often used to address white or grey secretions.

Furthermore, Calcium phosphoricum is recognized as the structural building block for bones and teeth, crucial for tissue growth and repair, and Kalium sulphuricum is tied to skin and mucous membrane function, playing a role in chronic inflammatory processes. These minerals are prepared in highly diluted (potentized) doses and administered as small, dissolvable tablets, underscoring the therapy's gentle, safe, and non-toxic nature.

Electrical Charge

One of the tenets of tissue salts is that the stable salts become unstable again and have an electrical charge once they begin to decompose, hence the first point of contact should be with the skin under the tongue (not with the hand). For example, Magnesium phosphate (Mag phos) begins to decompose in the saliva before reaching the acidic environment of the stomach, the necessary first step for the body to absorb its component minerals. Magnesium phosphate, with the formula Mg3(PO4)2, is an ionic salt and is generally insoluble in water. However, when it hits the stomach acid (hydrochloric acid, HCl), it reacts and breaks apart into its constituent ions: the metal Magnesium's positively charged ions (Mg2+) that the body absorbs these to use; and Phosphate ions (PO43−), which the body also absorbs. The compound continues to break down into soluble ions in the stomach and small intestine, and these ions are absorbed into the bloodstream.

So, Mag phos doesn't get absorbed as an intact molecule, but as a decomposition (Dissociation). The acidic conditions of the saliva and digestive system decompose it into charged ions so the minerals can be utilized by the body.

The philosophy of tissue salts–stressing balance and a holistic view of the body–finds striking philosophical harmony with ancient Eastern medical traditions, particularly Traditional Chinese Medicine (TCM) and Traditional Persian Medicine (TPM). In TCM, health is viewed as the coordinated function of an integrated whole, sustained by the smooth flow of vital forces like *Qi* and blood. The cellular mineral balance targeted by tissue salts can be seen as harmonizing the body's most subtle components, echoing the concept of *Jing* (Essence)–the foundational, refined substance that sustains life. Both therapies employ a keen, holistic diagnostic approach: Tissue Salts therapy observes physical symptoms like the color of secretions or the state of the tongue to pinpoint a mineral imbalance, which has similarities to TCM's diagnostic art of 'Looking, Listening, Questioning, and Palpating'. Crucially, both methods aim to enhance the body's innate self-healing capacity rather than merely targeting single symptoms.

Similarly, in Traditional Persian Medicine (TPM), the approach to health and disease is based on temperament (مزاج) and the balance of the four humors. Illness results from the deficiency or predominance of a temperament (hot, cold, wet, dry), and treatment aims to restore this equilibrium. This echoes the primary goal of tissue salts therapy: restoring the internal chemical balance. Its practitioners, like Avicenna and Rhazes, meticulously documented the therapeutic use of minerals and stones, classifying them by their temperament and specific action. For instance, their texts describe the use of substances like Hematite to stop bleeding, noting that, "Stones and salts, due to their dry and cold or hot and dry temperaments, are used in the treatment of wounds and injuries." This historical focus on the healing power of mineral substances, categorized by their elemental effects, bridges the ancient Eastern focus on refined substance and balance with the modern Western focus on microscopic cellular health.

Acceptance & Application

In today's global landscape, Tissue Salts Therapy is widely recognized as a valuable complementary medicine. It appeals to a growing number of people seeking gentle, supplementary support for daily health maintenance due to its established safety profile and lack of serious side effects. Because tissue salts do not typically conflict with pharmaceuticals, they serve as a flexible option for managing mild symptoms and enhancing overall vitality.

The pharmaceutical industry regards these as complementary therapies, citing the lack of documented scientific evidence equivalent to that which supports patented drugs. Such extensive tests could never find funding, since minerals cannot be patented and profited off of like modern pharmaceutical drugs. The money for expensive tests is reflected in the prices of these drugs. The medical industry stresses that tissue salts should never replace standard medical treatments for serious diseases.

Nevertheless, this growing, global interest in a therapy that adjusts body balance at the cellular level signals a broader trend: a desire to merge Western wisdom with Eastern traditional concepts for holistic human well-being. The drive for internal balance, though expressed differently, has a long historical precedent. Traditional Persian medicine, for instance, didn't use minerals in the modern cellular sense, but treated them as potent therapeutic agents, processing them into powders or solutions to balance the body's humors and temperaments. The focus was always on the chemical properties of natural substances to achieve specific physiological outcomes. A comparative analysis of these two therapeutic traditions–the modern cellular biochemistry of the West and the ancient elemental chemistry of the East–is now crucial to fully understand the deep-seated human focus on utilizing the body's own salts and natural materials for health.

A Comparative Analysis of the Ayurvedic Practice of Urine Distillation

The logic Dr. Schüßler used to derive tissue salts has its roots in the above ancient traditions, as well as Ayurvedic practices. A comparative analysis of these two distinct therapeutic traditions is in order, to juxtapose the 19th-century biochemic tissue salt therapy developed by Dr. Schüßler and the ancient Indian practice of urine distillation, with a specific focus on Gomutra (cow urine) and its distillate. While separated by millennia and cultural context, both systems are fundamentally rooted in the philosophical belief that maintaining the body's intrinsic balance of essential mineral substances is critical for health and that disease represents a deviation from this state. This report explores the historical origins, underlying principles, and modern scientific reception of each tradition.

Despite their shared philosophical foundation, the two practices diverge significantly in their methodology and, most importantly, their modern scientific standing. Schüßler's system, a form of homeopathic 'substitution therapy', utilizes infinitesimal doses of 12 inorganic salts to address specific, mineral deficiencies. Its efficacy is largely based on anecdotal reports and is not supported by rigorous scientific validation. The scientific and regulatory consensus, particularly from bodies like the U.S. Food and Drug Administration (FDA), regards these products as unapproved. In contrast, the traditional Indian practice of distilling Gomutra is not aimed at isolating a single salt but at creating a stable, potent liquid distillate (Kamdhenu Ark) believed to contain a complex array of bioactive compounds.

This practice has been the subject of some modern research, funded and conducted by agencies of the Indian government, including studies and US patents that have focused on its 'bioenhancing' properties. The proposed mechanism of action–enhancing the transport and effectiveness of conventional drugs–is a testable hypothesis that, while controversial, places the scientific inquiry into Gomutra in a distinct category from the unproven, highly diluted claims of tissue salts. This difference highlights a crucial divergence in the nature of their respective claims and the quality of their scientific engagement.

Bridging Two Worlds of Mineral-Based Therapies

The human quest for health and healing is a universal endeavor, often expressed through diverse cultural and historical lenses. Vast distances and epochs separate these two traditions: the biochemic system of medicine pioneered by the Schüßler in the 1870s, and the time-honored Indian tradition of urine therapy, or Shivambu, as documented in ancient Ayurvedic texts. It's worth deconstructing each practice and conducting a detailed comparative evaluation. While Schüßler's theory of biochemic tissue salts and the Ayurvedic use of urine distillation originated in different cultural and scientific contexts, both represent an enduring human effort to harness the restorative power of minerals and biological substances. A comparison reveals a shared philosophical ground—the knowledge of the body's innate mineral balance—but also starkly contrasting methodologies and profoundly different modern scientific evaluations.

Element or Mineral Salt?

An element is a pure substance consisting of only one type of atom. They are the fundamental building blocks of all matter.

Zinc is not a salt. Zinc is a chemical element and a type of metal. It has the chemical symbol Zn and an atomic number of 30. You can find zinc on the periodic table of elements. Metals like zinc are characterized by properties such as being good conductors of heat and electricity, being malleable and ductile, and having a shiny appearance. A common tissue salt made up of zinc is a compound called Zinc mur.

The Twelve Primary Tissue Salts

The twelve primary inorganic salts that Schüßler considered essential for cellular health are detailed in the following chart. The tradition behind this system attributes a wide range of therapeutic benefits to each of these historical salts, from treating colds and muscle aches to alleviating anxiety and depression. The following table provides a comprehensive list of these salts and their purported uses, as documented in the literature.

Mineral Name	Personality	Traditional Therapeutic Uses
(CaF_2) Calcarea fluorica or 'Calc fluor'	The elasticity salt	Helps strengthen tooth enamel & bones, restore tissue elasticity, relieve hemorrhoids & hernia pain
($Ca_3(PO_4)_2$) Calcarea phosphoric or 'Calc phos'	The bones & teeth salt	Helps restore cells & fractures; aids the digestive system
($CaSO_4$) Calcarea sulphurica or 'Calc sulph'	The cleansing processes salt	Helps purify blood, reduce infection, prevent sore throat & colds, treat skin disorders & acne
($Fe_3(PO_4)_2$) Ferrum phosphoricum 'Ferr phos'	The first assistance salt	Aids against inflammation; helps reduces fever, accelerates healing, reduces bleeding
(KCl) Kaliium muriaticum or 'Kali mur'	The mucous membrane salt	Helps purify blood, treats infection, reduces swelling, aids digestion
(K_3PO_4) Kalium phosphoricum or 'Kali phos'	The nerves and mind salt	Supports nerve health, lessens anxiety and fatigue, aids memory, relieves headaches

(K_2SO_4) Kalium sulphuricum or 'Kali sulph'	The chronic inflammation salt	Supports healing of mucous membranes and skin, balances metabolism
($Mg_3(PO_4)_2$) Magnesia phosphorica or 'Mag phos'	The cramps & pains salt	Helps relieve tension headaches & ease cramps, pain, and spasms
($NaCl$) Natrum muriaticum or 'Nat mur'	The fluid balance salt	Supports bodily fluid balance, reduction of water retention, aids digestion, eczema
(Na_3PO_4) Natrum phosphoricu m or 'Nat phos'	The acid-base salt	Supports relief of acidity, seasickness, arthritis, indigestion
(Na_2SO_4) Natrum sulphuricum or 'Nat sulph'	The excretion salt	Helps cleanse pancreas, kidneys, and liver; supports treatment of cold & flu
(SiO_2) Silicea or 'Silica'	The skin, hair & nails salt	Helps condition skin and connective tissue, cleanse blood, strengthen hair & nails

Disclaimer: This information is for educational purposes only and is not intended to diagnose, treat, cure, or prevent any disease. Always consult with a qualified healthcare professional before making any changes to your diet or health regimen.

These are just some of the essential elements necessary for human life. Natrum muriaticum is the homeopathic name for Sodium Chloride (NaCl), the chemical compound that makes up the majority of common table salt. An example of another essential element is zinc, which is found in the human body in the form of Zinc muriaticum. While zinc itself is not a salt, it is a key component in many different salts, such as:

- Zinc sulfate (ZnSO4)
- Zinc acetate (Zn(C 2H 3O 2) 2)

And the form found in tissues of the human body:

- Historically known as Zinc muriaticum, now called Zinc chloride, is ($ZnCl_2$)

These salt compounds are formed when zinc reacts with an acid. Zinc is a metal that acts like a base. As a metal, it is an element, and its chemical properties are generally described as a reducing agent (it readily loses electrons), like a base. These elements not only become more bioavailable when coupled as salts, but anecdotal descriptions portray them as being more effective when taken in certain combinations with other salts.

16

The Yoga (Combination) of Tissue Salts

"Seldom does a disease appear with the deficiency of a single cell salt, therefore, in the great majority of cases, it is necessary to administer a combination of two or more..."
— Boericke, W. and Dewey, W. A. (1914). *The Twelve Tissue Remedies of Schüßler*

Practitioners have found that salts work best in combinations of threes or fours and extended tissue salt lore to other areas of science and wellbeing. The word 'combination' itself, now used for these tissue salt remedies, and the Sanskrit word 'yoga' share a similar root meaning. 'Yoga' comes from the Sanskrit root 'yuj', which means 'to yoke', 'to join', or 'to unite'. The 'combination' it refers to is primarily the union of the mind, body, and spirit. In a broader, more philosophical context, yoga's meaning refers to the union of the individual self with universal consciousness or the divine. This is achieved through the practice of postures, breathing, and meditation, all of which are designed to bring the various parts of a person into harmony and alignment.

In a similar way, the 'combinations' of tissue salts listed in the document are about bringing together specific mineral salts to create a synergistic effect. Each individual mineral has been traditionally associated with a specific role in the body, but when combined in a particular formula (like Combination A associated with insomnia or Combination C associated with heartburn), their individual actions are amplified to address a specific ailment. This yoking of different elements for a common purpose is the fundamental connection between the two concepts. The tissue salts combine to restore balance and wellbeing to the body, much like the practice of yoga combines different aspects of a person to restore harmony. They help each other most in combinations of threes or fours. However, supplements with all twelve are reported to be too many, and they may undermine each other, since minerals can compete for absorption.

These recipes highlight the use of mineral salts commonly associated with traditional wellness and general health protocols, rather than treating specific medical conditions. For instance, combinations featuring Magnesium phosphate (Mag Phos) are traditionally known to support muscle function and are often used as part of a routine intended to assist the body's natural processes in maintaining connective tissues, which historically has been referenced in discussions about calcification or joint health concerns. Separately, Combination G, which includes Potassium phosphate (Kali phos), is frequently utilized in traditional practices for its potential role in promoting a sense of nervous system health and emotional balance, a quality some have associated with supporting comfort from conditions like back ache. Finally, a recipe like Combination J is often incorporated into traditional protocols for general respiratory comfort and may be used to address minor seasonal symptoms.

These combinations are taken on an empty stomach, and left to absorb under the tongue. Many people get all 12 of them in separate bottles, so they can mix them in different combinations according to the ailment, starting by getting the relevant ones first, but individually. They are also sold as combinations (associated ailments below):

Combination A

- Ailments: Insomnia, sciatica, neuralgia, cramps, neuritis, nervous tension,

restlessness, cramps, inflammation
- Ingredients: FERR PHOS, KALI PHOS, MAG PHOS

Combination B

- Ailments: Irritability, anxiety, general debility, nervous exhaustion, convalescence, fatigue, stress
- Ingredients: CALC PHOS, KALI PHOS, FERR PHOS

Combination C

- Ailments: acidity, heartburn, dyspepsia, acid reflux, indigestion
- Ingredients: MAG PHOS, NAT PHOS, NAT SULPH, SILICA

Combination D

- Ailments: skin conditions including dermatitis, acne, pimples, boils, eczema, glandular swelling, tonsillitis & an ulcerated sore throat.
- Ingredients: KALI MUR, KALI SULPH, CALC SULPH, SILICA

Combination E

- Ailments: indigestion, colic, flatulence, stomach cramps, nausea
- Ingredients: CALC PHOS, MAG PHOS, NAT PHOS, NAT SULPH

Combination F

- Ailments: Fatigue, Migraine, Headache, Nervous Headache, Nervous tension, mental exhaustion
- Ingredients: KALI PHOS, MAG PHOS, NAT MUR, SILICA

Combination G

- Ailments: Lumbago, Backache, Hemorrhoids/Piles, Loss of elasticity, Weak connective tissue, varicose veins
- Ingredients: CALC FLUOR, CALC PHOS, KALI PHOS, NAT MUR

Combination H

- Ailments: Allergic Rhinitis, Hay fever, Runny nose, sneezing, itchy eyes
- Ingredients: MAG PHOS, NAT MUR, SILICA

Combination I

- Ailments: Muscular pain, Fibrositis, Stiff neck, aching joints, cramps
- Ingredients: FERR PHOS, KALI SULPH, MAG PHOS

Combination J

- Ailments: Coughs, Colds, Congestion, Sore throat, chest congestion, feverish colds
- Ingredients: FERR PHOS, KALI MUR, NAT MUR

Combination K

- Ailments: Weak & brittle nails, falling hair, Brittle bones, lack of vitality
- Ingredients: KALI SULPH, NAT MUR, SILICA

Combination L

- Ailments: Circulatory disorders, varicose veins, Hardened arteries, poor blood flow
- Ingredients: CALC FLUOR, FERR PHOS, NAT MUR

Combination M

- Ailments: Rheumatism, Joint pain, swelling, muscle soreness
- Ingredients: CALC PHOS, KALI MUR, NAT PHOS, NAT SULPH

Combination N

- Ailments: Menstrual Pain, Period cramps, painful periods
- Ingredients: CALC PHOS, KALI MUR, KALI PHOS, MAG PHOS

Combination P

- Ailments: Poor circulation, chilblains, aching feet and legs, Cold hands and feet, numbness
- Ingredients: CALC FLUOR, CALC PHOS, KALI PHOS, MAG PHOS

Combination Q

- Ailments: Sinus, Catarrh, Headaches from congestion, runny nose
- Ingredients: FERR PHOS, KALI MUR, KALI SULPH, NAT MUR

Combination R

- Ailments: Erupting teeth, infants' teething pain, Restlessness, irritability, swollen gums
- Ingredients: CALC FLUOR, CALC PHOS, FERR PHOS, MAG PHOS, SILICA

Combination S

- Ailments: Digestive disorders, Stomach complaints, Biliousness, Sick headache, Sluggish liver, yellow-coated tongue
- Ingredients: KALI MUR, NAT PHOS, NAT SULPH

Combination T

- Ailments: Inflammation, fever, illness, chills
- Ingredients: FERR PHOS, KALI MUR

Combination U

- Ailments: Poor calcium absorption, weak bones, poor nutrient assimilation
- Ingredients: CALC FLUOR, CALC PHOS, NAT PHOS, SILICA

Combination 5 (Nerve Tonic)

- Ailments: Nerve Tonic, Anxiety, weakness, neuralgic pain
- Ingredients: CALC PHOS, FERR PHOS, KALI PHOS, MAG PHOS, NAT PHOS

Combination 12 (General Tonic)

- Ailments: General Tonic, Debility, fatigue, exhaustion
- Ingredients: CALC FLUOR, CALC PHOS, FERR PHOS, KALI MUR, KALI PHOS, KALI SULPH, MAG PHOS, NAT MUR, NAT PHOS, NAT SULPH, SILICA, CALC SULPH

Note that the specific designation of 'Combination O' does not appear to be a standardized name in the same way that the above combinations are, suggesting that homeopaths can round out this table. In fact, the naming conventions for these cell salt combinations (such as 'A' through 'Z') can vary significantly between different homeopathic practitioners, manufacturers, and books. For example, the missing Combination O, could be made from combinations not represented above. A combination of NAT SULPH and SILICA, might be a plausible and logical pairing for the ailments oily skin, metabolic sluggishness:

Combination O

- Ailments: Oily skin, oily hair, obesity, glandular swelling, metabolic sluggishness
- Ingredients: NAT SULPH, SILICA

Since tissue salts are sold in combinations, or separately, the ones sold separately can be used to make up these combinations, usually of 3, sometimes 4, salts. Possible combinations to round out the above table might be:

Combination V

Ailments: Constipation, Bloating, Dryness (skin, hair, mucous membranes), Water retention

Ingredients: NAT MUR, NAT SULPH,

SILICA

Selecting a logical combination of the Schüßler's Cell Salts (like NAT MUR, NAT SULPH, and SILICA), homeopaths can deduce the ailments that those specific ingredients are known to treat in combination. For example, since those three salts are all related to fluid balance and detoxification, the list of ailments for 'Combination V' was a direct result of their known functions. This combination was designed to address the listed ailments based on the known functions of each ingredient:

NAT MUR is used for balancing moisture in the body, addressing both dryness (constipation) and excess water retention (bloating).

NAT SULPH helps with fluid elimination and liver function, which can improve digestion and reduce bloating.

SILICA is a deep-acting remedy for poor assimilation and helps the body with the elimination of waste, supporting relief from chronic constipation.

Combination W

- Ailments: Feverish conditions, Chills, Heat flashes, Fever, Acute inflammation, Body aches
- Ingredients: FERR PHOS, KALI MUR, NAT SULPH

Combination X

- Ailments: Circulatory problems, Anaemia, Cold hands and feet, Paleness, Numbness, Lack of vitality
- Ingredients: FERR PHOS, KALI MUR, NAT MUR

Combination Y

- Ailments: Respiratory issues, Asthma, Bronchitis, wheezing, rattling cough, chest congestion
- Ingredients: FERR PHOS, KALI MUR, NAT SULPH, MAG PHOS

Combination Z

- Ailments: Sinusitis, Catarrh, Post-nasal drip, Sneezing, Runny nose, Stuffy nose
- Ingredients: KALI MUR, KALI SULPH, NAT MUR, NAT PHOS

Ingredients can be chosen based on their established functions:

NAT MUR (Sodium Chloride) balances moisture in the body, which addresses both dryness (constipation) and excessive water retention (bloating).

NAT SULPH (Sodium Sulphate) helps with fluid elimination and liver function, which can improve digestion and reduce bloating.

SILICA (Silicon Dioxide) is a deep-acting salt that helps with the elimination of waste and supports the body's overall cleansing process, making it a good fit for chronic constipation.

Nuances must be observed. Whereas Mg (Magnesium) can cause muscle cramps, magnesium phosphate (Mag phos) is purported to relieve them. Regarding the efficacy of diluted homeopathic remedies, more concentrated solutions of the same tissue salts are available: 6X (diluted six times), as opposed to 12X (diluted 12 times). The traditional rationale is that smaller doses are in order, since the versions of the salts found in human tissues are more bioavailable than minerals in forms other than those found in the human body.

Each of these salts is traditionally associated with a different set of symptoms or bodily function, from supporting bone and nerve health to treating inflammation and skin issues. The minerals gravitate toward the relevant organ because the body has specific cells for specific organs and tissues. The difference in the cells is largely determined by the type and composition of the inorganic salts that make up a cell.

Tissue salts uphold a natural approach to healthy living. The reimbursement of these simple mineral messengers (tissue salts) stands as a litmus test for a nation's philosophical commitment to holistic health. While the financial titans of allopathy dominate global coverage, a few countries bravely step into the realm of natural efficacy: India leads the charge, enshrining these remedies under the expansive AYUSH mandate, granting them governmental legitimacy and broad access.

In the heart of Europe, Germany offers a lifeline, with its private insurers fully embracing consultations and statutory funds providing optional benefits, defying the conventional skepticism. Nearby, Switzerland maintains a delicate balance, periodically re-embracing them in its basic health plan, recognizing the public's demand for choice.

Even smaller nations like Belgium offer partial public coverage of tissue salts, while the United Kingdom, Austria, Hungary, and the Netherlands rely on the private insurance sector to bridge the gap, proving that where government regulation hesitates, the people's desire for natural healing still finds an economic path. A testament to the fact that true health begins not with a new drug, but with the institutional courage to financially legitimize personal responsibility. Accessible, natural care shows that the simplicity of fundamental wellness is valued more highly than the complexity of engineered disease.

17

Patentability VS. Ancient Roots

"...these remedies must be given in small doses..... Every biochemical remedy must be thus attenuated, so that the functions of the healthy cells may not be disturbed, and yet the functional disturbances present may be equalized."
— Dr. Med. Schüßler, *An abridged therapy manual for the biochemical treatment of disease* (1898).

Bioavailability, the elegant name for the body's assimilation of a substance, is a variable feat. While many salts, a vast and varied class of compounds, are indeed consumed, their absorption is a clumsy process. The body, a connoisseur of what it can use, often rejects or struggles to utilize these foreign substances. Tissue salts, in contrast, are the pristine compound, a variety of satisfactions perfectly suited to our biology. These are the very salts the human body is made of. For example, the use of Magnesium phosphate, as opposed to common Magnesium citrate, results in a kind of biological harmony, a seamless integration into the cellular fabric. The body's ability to use these substances is an elegant dance between compound and cell, making them significantly more effective than their cruder, less-refined counterparts.

The Un-Patentable 'Drug'

The modern scientific consensus on biochemic tissue salt therapy and homeopathy at large uses as a crutch this supposed 'lack of rigorous, independently verified evidence' to support its claims of effectiveness. That argument backfires to expose the Achilles' heel of drug testing. Virtually all multi-billion dollar drug tests are conducted on items that pharmaceutical companies can patent. The pharmaceutical industry's business model is built on this myopic principle.

Dr. Servan-Schreiber's book, *A New Way of Life,* describes how the mainstream medical establishment is not doing research on the health benefits of food. Pharmaceutical science, a vast and varied practice, finds its satisfaction in the singular: the isolation of a molecule, its synthesis, and a profitable patent. The pharmaceutical industry is myopic in its focus on isolating a single active molecule from a plant or food source, synthesizing it, and then patenting that specific chemical compound. A masterpiece, like Taxol from the humble yew, is a rarity in this market of chemical proliferation. This is why you see drugs derived from plants like Taxol from the yew tree, but not the plants themselves being sold as a medical treatment.

The immense cost of drug research and development, which can run into the hundreds of millions, to billions of dollars per drug, is only financially viable because a successful patent grants the company a temporary monopoly on the drug's sales. This allows them to recoup their $4 billion investment and generate a profit. The result is that drug companies use the 'untested' and 'unproven' argument against food in favor of chemicals, as explained in Dr. David Servan-Schreiber's seminal book, *Anticancer: A New Way of Life*.

Clinical trials required to bring a new drug to market can cost billions of dollars, an investment is only made by pharmaceutical companies because they can patent the drug and recoup their costs, along with a significant profit, once it's approved. This vast and complex system, with its ever-proliferating intellectual property, has set the pharmaceutical industry in direct competition with a much more elemental source of health: food itself.

Foods and other natural compounds, by their nature, cannot be patented. Therefore, there is no financial incentive for a company to spend a vast amount of money to prove that a specific food, like broccoli or turmeric, has an anticancer effect. In this system of commercial genius, the raw materials of nature—the broccoli and turmeric of the world—are seen as a retrograde feat. Unpatentable and thus unprofitable, there is no incentive for anyone to pursue the grand, naive gesture of proving their efficacy. These are the humble materials of an unremunerated art, unworthy of a master's attention.

Common sense tells us we can't survive on chemicals alone; we need food to live. Without the kind of funding only a government could afford, the lack of research into food's medicinal properties—a deficit so widespread it's almost an art form—is a red flag in itself.

In fact, the inability to test foods, to truly know their healing power, is a simple economic reality. The primary reason foods are not tested for their medicinal properties is a fundamental economic one: Private industry's inability to patent a food item, make the research money back, and turn a profit. Conversely, the requirement to spend exorbitant amounts of money on drug testing, the proliferation of regulations and bureaucratic rituals, is a testament to the fact that these drugs are fundamentally more expensive, and in many cases more harmful than food.

Against this backdrop, the broader work of healing–that of the food itself–is dismissed as a retrograde feat, a kind of naive, holistic fantasy. Servan-Schreiber argues for an individual's responsibility in the face of this devaluation, a personal embrace of a diet rich in foods with anecdotally proven anticancer properties.

Hence, individuals must take control of their health by adopting a 'new way of life'. People need to incorporate a diet rich in foods that have been shown anecdotally or in small studies to have anticancer properties. Ultimately, people need to be aware of the value the government places on patented molecules over—and to the detriment of—a patient's self-discipline, essentially legislating a preference for profit over genuine, self-sustaining wellness.

Ayurveda

Indian medicine, most powerfully embodied by Ayurveda is a testament to the profound power of the wild and government responsibility regarding health. This practice of healing stands in stark contrast to the Western chemical arsenal, for studies on Ayurveda have been funded and conducted by agencies of the Indian government, and Ayurveda is covered and reimbursed by most major health insurance policies in India.

Drawing its entire strength from remedies and rituals derived directly from nature's untouched bounty, this ancient science posits that the true path to wellness is found in Earth's garden, minimizing the need for the synthetic compounds and chemically-developed abstractions that define contemporary pharmaceuticals. The treatments are a meticulous symphony of plant life, composed predominantly of herbs, spices, and fruits—like the golden strength of Turmeric or the restorative power of Ashwagandha—often prepared with a reverence that keeps them as close as possible to their original, potent form. This deliberate rejection of over-processing is key, offering a gentle yet powerful intervention intended to work with the body, not violently against it.

Ancient Roots and Cultural Significance

Within Ayurvedic medicine, there is a practice that evokes a logic similar to that of

Dr. Schüßler's organ burning experiment, albeit a tad more straightforward. The Indian traditional distillatio practice of urine therapy, or Shivambu, is a time-honored tradition deeply rooted in Ayurvedic medicine and Hindu and Tantric religious traditions.

In ancient times, urine was at times not viewed as a waste product to be discarded, but rather as a distilled product containing useful substances for healing the body. It was referred to in sacred texts as the 'gold of the blood' and the 'elixir of long life', indicating its therapeutic and sacred status. This practice, which is attributed to Indian culture, has a long history, with records of its use found in cultures from the Egyptians and Greeks to the Romans, continuing through the Middle Ages and into the modern era.

Minerals Distilled from Urine

In the Ayurvedic tradition, cow urine (Gomutra) holds a particularly prominent position. It is extensively described in ancient Ayurvedic texts like the Sushruta Samhita and Ashtanga Sangraha, which classify it as an effective medicinal substance. *Bhav Prakash Nighantu,* another key Ayurvedic text, describes Gomutra as the most superior of all animal urines and a key component of the mixture known as Panchagavya.

The therapeutic properties of Gomutra are attributed to its complex chemical composition. Modern biochemical analyses have shown that cow urine is approximately 95% water, with the remaining 5% consisting of a mixture of urea, minerals, salts, hormones, and enzymes. It contains a wide array of compounds, including sodium, nitrogen, sulfur, vitamins, manganese, iron, silicon, chlorine, magnesium, calcium salts, and a variety of acids and enzymes.

The Process of Distillation

The practice of distilling salt from urine in

the context of traditional Indian medicine, the primary purpose of distillation is not to extract solid salt for consumption. Instead, the process is used to create a stable, more potent liquid distillate known as Kamdhenu Ark. The distillation process is believed to enhance the qualities of the cow urine, making it more effective for therapeutic purposes. This process can even be performed at home using simple equipment like a pressure cooker.

While the traditional therapeutic practice focuses on creating a liquid distillate, other sources mention alternative, more modern methods for extracting salt from urine. For example, a method for gardeners involves simply placing urine in a refrigerator to precipitate out salt and urea, which adhere to the container walls.

Historically, chemists like Hilaire Rouelle and Carl Wilhelm Scheele developed a more complex process to isolate urea crystals by treating urine with nitric acid and alcohol. This distinction is critical: the traditional therapeutic practice of distillation is a method for purification and concentration, while other methods are designed for chemical extraction of specific compounds like solid salt or urea for non-therapeutic purposes.

Modern Research and Scientific Claims

Unlike tissue salts, which have been largely dismissed by the scientific community, the therapeutic claims of Gomutra have been subjected to a different type of scientific inquiry. This is the most crucial point of distinction between the two traditions. Modern research has focused on the 'bioenhancing' properties of the cow urine distillate.

A bioenhancer is a substance that increases the effectiveness and bioavailability of other drugs. Studies have been conducted showing that cow urine distillate is a more effective bioenhancer than the unprocessed urine itself. The proposed mechanism is that it enhances the transport of various antibiotics, such as Rifampicin, Tetracycline, and Ampicillin, across the gut wall and cell membranes. One study found that it increased the action of Rifampicin against Escherichia coli up to sevenfold and against Gram-positive bacteria up to 11-fold.

Beyond its bioenhancing capacity, research snippets also detail other specific claims, such as its antimicrobial properties due to the presence of urea, and its anticancer effects attributed to the antioxidant properties of uric acid and allantoin. The existence of US Patents (No. 6,896,907 and 6,410,059) for the medicinal properties of cow urine, particularly as a bioenhancer and as an antibiotic and anticancer agent, elevates the discussion beyond purely anecdotal claims and places it within the framework of modern intellectual property and scientific inquiry.

The focus on the bioenhancing function represents a profound conceptual shift from the all-encompassing claims of ancient traditions. It re-frames the substance not as a standalone cure-all, but as a potentiating agent that can improve the efficacy of existing conventional drug therapies. This claim is far more specific, testable, and scientifically-oriented than the vague assertions of tissue salts, making the modern scientific conversation around Gomutra fundamentally different.

Ultimately, the natural spirit of Indian medicine is revealed in its unflinching, government backed focus on prevention and the root cause. It is a system built not around the dramatic, last-minute rescue, but around the slow, deliberate cultivation of the body's innate power to heal itself. The primary campaign is to preempt the shadow of illness by diligently preserving a balanced mind, body, and spirit through righteous diet and mindful living. Instead of the superficial suppression of visible distress, Ayurvedic treatments embark on an arduous journey to confront the deep-seated imbalance that first opened the door to disease. This profound dedication to total restoration culminates in intense cleansing rituals, such as Panchakarma, a comprehensive detoxification designed to purge the treacherous accumulation of internal toxins (Ama) and secure a truly natural, sustainable, and lasting victory over affliction.

A Comparative and Shared Ground

Despite their disparate origins, both Schüßler's biochemic system and the Ayurvedic tradition of urine therapy operate on a common philosophical principle. Both are built on the foundation that health is a state of internal equilibrium of essential substances–be they 12 specific mineral salts or a complex mixture of compounds in a bodily fluid. This perspective views illness not as a foreign invader but as an internal imbalance or deficiency that can be corrected by introducing the missing or deficient substance. This shared philosophy reflects a pre-modern understanding of health as a state of harmony and balance, a concept that persists in many traditional and alternative medical systems today.

Contrasting Scientific Community Reception

The two traditions diverge sharply in their practical application. Schüßler's system is a targeted, single-compound therapy that utilizes a specific, rigid list of 12 inorganic salts. The methodology relies on a micro-dosing approach, with each salt administered to address a specific, identified ailment. The system's success, according to its proponents, is in the correct identification of the single deficient salt.

Conversely, the Gomutra tradition is not about administering a single, isolated compound. The practice involves using a complex, multi-compound liquid distillate (Kamdhenu Ark) that contains a wide variety of substances acting in concert. This approach reflects a more ancient, holistic view of healing, where a single therapeutic agent is believed to have multiple, synergistic effects on the body.

The most significant distinction between the two traditions lies in their reception by the modern scientific community and regulatory bodies. The biochemic system of medicine, due to its reliance on homeopathic principles of extreme dilution and anecdotal evidence, is the less proven of the two. Rather than researching Schüßler's claims, the scientific community in the West rests on the argument that it sees no evidence that tissue salt preparations are effective. The claims of tissue salts are supposedly not falsifiable within the framework of modern medicine because the concentration of the active ingredient is often too low to elicit a measurable pharmacological response.

The discourse surrounding Gomutra is more nuanced. While still considered an alternative medicine and not a part of mainstream medical practice, it has been the subject of research funded and conducted by agencies of the Indian government that has produced specific, testable, and scientifically framed claims. The focus on its 'bioenhancing' properties and its ability to improve drug transport across membranes provides a concrete, plausible, and falsifiable hypothesis that can be investigated with modern research methods. Furthermore, the existence of US patents related to its properties gives it a unique standing, as it has been subjected to a formal legal and scientific review process that is far more rigorous, thanks to the research agencies of the Indian government, than a reliance on anecdotal reports. This places Gomutra in a different category from the discredited claims of homeopathy, as its value is being explored as a potential adjunct to conventional medicine rather than a replacement.

While tissue salts are a simple, single-compound cure administered in minimal doses according to the 'law of minimum dose', Gomutra is a more concentrated liquid distillate employed as a complex, multi-compound agent.

Two Paths, One Quest

In conclusion, the discovery of Schüßler's biochemic tissue salts and the ancient Indian tradition of urine distillation represent two distinct but spiritually related responses to the fundamental human desire for health and healing. Both traditions are built on the philosophical premise that restoring the body's intrinsic, bioidentical mineral and chemical balance is the key to overcoming disease. This shared belief reflects a universal quest to find accessible remedies in natural substances, a pursuit that has been at the core of human health practices for millennia.

But the ancient practice of urine distillation

in India, while controversial and rooted in a pre-scientific tradition, has been the subject of some modern research that has yielded specific and testable claims. The focus on its 'bioenhancing' properties–its ability to increase the efficacy of conventional drugs–is a scientifically-oriented hypothesis that has led to intellectual property recognition, a distinction not afforded to homeopathy, where tissue salts have been pigeon-holed.

This reframes the substance from a mystical elixir to a potential adjunct therapy, a testable claim that can be verified or refuted through further scientific investigation.

Ultimately, while both traditions hold significant cultural and historical importance, the modern scientific inquiry into each has led to vastly different conclusions about their potential value and credibility. This analysis underscores the importance of distinguishing between publicly and privately funded research, and scientifically validated treatments.

Lessons Learned

The fact that the Indian government funded and conducted research on natural medicine and food shows that it is not a system that profits more from managing illness than curing it, unlike some others that have implicitly declared a preference for customers over citizens. When policymakers confuse health with pharmacological dependence, they trade the robust independence of nature for the fragility of a chemical prison. The institutions a society legitimizes define the boundaries of its freedom. When the state, through its regulatory power, grants financial primacy to chemical intervention, it sets the example with a precedent that dictates not just the terms of health, but the very nature of life itself—a life dependent on continuous industrial output rather than innate, self-sustaining wisdom.

The Professional Paradox

The professional paradox reveals that a society's indoctrinated healers can become the unwitting custodians of its chronic malaise. Whereas Indian medicine aims for the symbiotic relationship between government regulation (insurance mandates), pharmaceutical companies, and the medical profession, the Western system upholds the damage it purports to fix. Without getting into scandals around organ harvesting, at least critics admit that the Western system is designed to manage chronic disease rather than cure it: by financially favoring patented chemical treatments (drugs), a cycle is created where the 'fix' (a pill) leads to side effects, requiring more chemical or even surgical intervention, thus causing the very damage they purport to fix–a self-perpetuating system of dependency and increasing medical cost.

The Government sets the tone. When a government's insurance structure overwhelmingly mandates coverage for high-cost, interventionist chemical therapies while requiring patients to pay cash for preventive, lifestyle, or traditional/natural care (like minerals, yoga, acupuncture, nutritionists), it signals to industry and the populace that health is a commodity delivered through a pill, not a state achieved through food (including minerals), diet, discipline and self-care. The state's greatest tyranny is the subtle economic direction that leads the public away from innate wisdom and toward industrial dependency.

18

Salt Outmoded

"The stinging salt... is the role of poets and writers—to preserve the body of society from decay." — Vilhelm Moberg (1898-1973)

You might have heard grandparents still using the term 'icebox' interchangeably with 'freezer'. It's hard to imagine the time when salting, drying, and smoking were the most common ways to preserve food.

While some people used natural ice stored in ice houses, this was unreliable and often a luxury. Refrigeration replaced salt as a primary food preservative on a large scale over a period from the late 19th to the mid-20th century. This shift was driven by the invention and mass production of mechanical refrigerators, which fundamentally changed how food was stored and distributed on every level of the supply chain.

The Era of Mechanical Refrigeration

The first refrigeration machines of the late 1800s were not the work of a single inventor, but were made commercially viable through the contributions of several engineers, most notably the German engineer, Carl von Linde. While the core principle of vapor-compression refrigeration was established earlier (by people like Jacob Perkins and John Gorrie), Linde's key contributions in the 1870s and 1880s were crucial for the industrial scale-up mentioned in the late 1800s.

In 1876, Linde patented an improved and more efficient process for liquefying gases, which led to the creation of the first reliable and efficient compressed-ammonia refrigerator. This breakthrough was essential because it allowed the new mechanical refrigeration technology to be widely adopted by industries like breweries and meatpacking plants, ushering in the age of commercial refrigeration. These were large and expensive, so they were first used in commercial industries, such as breweries and meatpacking plants. This innovation allowed these industries to produce and transport goods without relying on huge amounts of salt.

From the 1920s-1930s, the household electric refrigerator began to see widespread adoption in the United States. With the development of more compact, safer, and affordable units, refrigerators moved from being a luxury item to a common appliance in middle-class homes. This provided a convenient and effective way for families to keep food fresh for days or weeks.

By mid-20th century, the rapid increase in household refrigeration, combined with the development of refrigerated transportation (like refrigerated rail cars and trucks), created a 'cold chain' that revolutionized the food supply. This system allowed for fresh produce, meat, and dairy to be transported across the country and sold year-round, which significantly reduced the need for heavy salting for preservation. While salt is still used today in processed foods for flavor and some minor preservation, it no longer holds its historical role as the sole, essential method for keeping food from spoiling.

The Meatpacking Industry Revolution

Before refrigerators replaced salting, meat was heavily salted and preserved for shipping. The meatpacking industry was one of the first to be transformed by refrigeration, which made it possible to centralize production and ship fresh meat nationwide. The shift from salting to refrigeration was a massive cultural change, and it led to some funny moments and anecdotes as people adjusted to this new technology. The humor often came from the clash between old habits and new conveniences.

Chicago's rise as a meatpacking hub was largely due to the development of the refrigerated railway car. This innovation allowed fresh meat to be transported from the Midwest to the East Coast, eliminating the need to ship live animals or heavily salted meat. This directly replaced salting as the primary method of meat preservation for long-distance transport. *Chicago and the Great West,* by William Cronon, details how refrigeration and the railroad network fundamentally restructured the American economy and the relationship between cities and their hinterlands, with Chicago serving as the central hub.

People were at a loss to keep up with this fast-paced innovation. A classic trope from the early days of refrigeration was the older generation, used to a lifetime of salting and preserving, who simply couldn't trust the newfangled electric box. A great-grandmother who was a master of cured ham, a legend in the family for her smoked meats that would last a year in the cellar, would turn skeptic when their son brought home their first refrigerator in the 1930s. She would just stare at it, ignoring promises like, "Mom, this will keep food fresh for days, even weeks!" She would shake her head and say, "Hmph. You mean to tell me a little electric fan can do what it takes a whole barrel of salt and a month of patience to do?" He would find her a week later, quietly trying to sneak a pound of salt-cured bacon into the fridge, just to be safe—not taking any chances with the family's provisions.

The Shift in Fisheries and Seafood Transport

Before refrigeration, fresh fish was a delicacy limited to coastal communities. Salting was the only way to preserve it for inland markets. The invention of the refrigerated fishing vessel and cold storage facilities in port cities allowed for fresh fish to be distributed far inland. This development made seafood a staple in places like the American Midwest, where previously only heavily salted or dried fish was available.

The Household Refrigerator & Everyday Life

The widespread adoption of the home refrigerator in the early to mid-20th century directly changed family food preservation habits. In the 1920s and 1930s, advertisements for electric refrigerators often highlighted the end of salting and the ability to keep food 'fresh'.

The shift was a major selling point, liberating homemakers from the labor-intensive process of curing and brining. The refrigerator made it possible to buy food in larger quantities without fear of it spoiling, changing shopping habits from daily trips to weekly ones. A real historical trend–the cultural skepticism of new technology–provides a humorous light and reference for the underlying phenomenon.

Competing with the 'Electric Icebox'

The transition to home refrigerators was met with a degree of skepticism, particularly in rural areas where people were accustomed to time-honored preservation methods like salting, drying, and iceboxes. Farmers, who often had a deep understanding of food preservation, were particularly hesitant to trust a new, complex machine. The humor lies in the clash between traditional knowledge and modern convenience.

A common theme in historical advertising and oral histories was the initial suspicion toward the refrigerator. People questioned how an 'electric box' could possibly be more effective than a century-old barrel of salt. The joke was often on the machine, which had to prove its worth.

The introduction of household appliances met with cultural resistance and skepticism. People found it humorous to be 'outsmarted' by a machine, especially when they had perfected a skill like salting over generations.

The Advent of the Overstuffed Refrigerator

As refrigerators became more common, people didn't immediately change their buying habits. The humor came from the paradox of a new appliance being filled to the brim, with a mix of fresh food and old preserved items.

W. Schivelbusch's *Tastes of Paradise: A Social History of Spices, Stimulants, and Intoxicants* describes how an early owner of a refrigerator might have continued to can and salt items out of habit or as a backup. The refrigerator, a symbol of freshness, would become a jumbled repository for everything from a fresh carton of milk to a jar of heavily salted pickles made the year before. The cultural lag resonates today, with new technology being used in old ways. The idea of the 'overstuffed' refrigerator can be linked to the broader trend of abundance and a shift from a preservation-based diet to a freshness-based one, which was a source of both pride and a bit of comical overload for early adopters. Industrialization's teen years were full of awkwardness, in an era rooted in the cultural friction between established practices and new, disruptive technology.

All this talk of refrigeration makes one wonder whether salt was also replaced in other sectors, like medicine, by pharmaceuticals. Not really. Half of pharma is salt. People still buy Epsom salt and soak in the magnesium in hot winter baths, and they still 'recharge their batteries' swimming in the salty sea.

19

The Osmosis of Swimming in Salts

"...the mightiest animated mass that has survived the flood; most monstrous and most mountainous! That Himmalehan, salt-sea Mastodon, clothed with such portentousness of unconscious power..."
—Herman Melville, *Moby Dick*

The 'electric charge'–that peculiar, pleasant tingling swimmers report after months of swimming in the sea–is deep physiological shift. The sensation, described as a tingly feeling, may be an individual's way of translating the body's reaction to several powerful forces at work. One theory involves grounding, suggesting that direct contact with the sea's conductive water allows the body to absorb free electrons, generating a subtle, antioxidant-driven feeling of wellbeing that a swimmer might interpret as an uplifting, electric exchange. The 'charge' is a saltwater phenomenon, not one resulting from chlorine pools, so more likely fueled by the massive endorphin release triggered by the immersion in saltwater. This hormonal flood of natural 'feel-good' neuropeptide hormones induces a state of euphoria and alertness, creating an overwhelming sense of being powerfully 'charged' or 'electrically alive' by the experience itself.

The skin, far from being a simple covering, is a colossal organ—the body's largest—a fortress of layered tissue functioning as the primary shield of the integumentary system. This classification is earned by the collective and coordinated effort of its three distinct structural layers: the epidermis, the dermis, and the hypodermis. The outermost Epidermis forms the regenerating, waterproof frontier, armed with keratinocytes for protection and melanocytes for defense against the sun's radiation. Beneath this lies the muscular dermis, a thick, resilient matrix laced with the strong fibers of collagen and the elastic snap of elastin, providing the organ with its strength and flexibility. It is here that the intricate network of nerve endings registers every sensation, while sweat and oil glands meticulously manage temperature and moisture. Finally, the deepest layer, the hypodermis (subcutaneous layer), acts as a protective, fatty cushion, serving as insulation against the cold and a dynamic reserve for energy storage. Only by integrating these tissues and their specialized cells–from immune Langerhans cells to temperature-regulating glands–can the skin successfully execute the essential functions of protection, sensation, and regulation, confirming its vital status as a complex organ.

The skin's status as a complex organ is intrinsically tied to its principal function: acting as a formidable barrier. This protective role fundamentally dictates its interaction with environmental minerals, setting up a clear contrast between what the body is meant to keep out and what it's meant to absorb for its own health. The primary structure governing mineral entry is the epidermis, specifically the lipid-rich stratum corneum. This protective layer is exceptionally effective against highly charged, water-soluble ions like sodium and chloride, such as those found in seawater.

When the skin is exposed to concentrated saltwater, instead of mineral absorption, the skin experiences osmosis, losing water to the concentrated external environment. The primary function here is exclusion, reinforcing the skin's role as a protective organ, not an absorptive one. To be systemically absorbed, the vast majority of large, water-soluble, inorganic mineral compounds cannot bypass the skin's tough lipid-based barrier, meaning the body reliably relies on the digestive system for its mineral intake.

Despite the skin's excellent barrier function, small, temporary channels allow for the minor, localized absorption of a few specific minerals, creating a few exceptions to the rule. Magnesium is the most cited of these exceptions, with evidence suggesting its ions can utilize the tiny bypass channels of hair follicles and sweat glands to penetrate the epidermis and reach the dermis, where they are picked up by the extensive network of blood vessels. This effect is strongly dependent on high concentration (such as in Epsom salts or mineral pools) and duration of exposure.

Trace amounts of other minerals, such as calcium and potassium, may also enter through these secondary pathways, but the effect is predominantly topical, benefiting the skin's surface cell functions like hydration and barrier repair rather than contributing significantly to the body's overall mineral balance.

The interaction of the skin with chemicals in a chlorine pool represents a chemical attack on the organ's barrier function, distinguishing it from beneficial mineral absorption. Chlorine is a powerful oxidizing agent designed to destroy organic material, and when it contacts the skin, it is absorbed not for nourishment, but due to damage. Chlorine actively strips away the essential oils and lipids–the protective matrix of the stratum corneum–thereby compromising the skin's barrier. Because this barrier is chemically compromised, chlorine (and its byproducts, chloramines) can be absorbed through the damaged skin. This absorption route is exacerbated by the fact that the primary exposure routes in a pool are dermal absorption (accelerated by hot water) and, more significantly in indoor pools, inhalation of the chemical vapors. In summary, the skin's function as a complex organ is defined by its selective barrier: it successfully repels most therapeutic minerals from the environment, but it is vulnerable to the destructive chemical properties of chlorine, which temporarily compromises its protective structure to gain entry.

Swimming is excellent exercise, but its health benefits vary depending on the water you're in. While a chlorine-treated pool offers a consistent workout environment with minimal risk of pathogens, not only is it unlikely to give you a 'charge', but it can also lead to skin, eye, and even organ irritation. Swimming in natural bodies of water offers a unique experience and provides therapeutic benefits from its high mineral salt content, as long as you are mindful of currents, marine life, and pollution.

The Chlorine in Pools: A Salt or Not?

The chlorine in pools is not a salt itself, but it is often derived from salts. Although chlorine is not a salt, it is a fundamental component of many salts. Chlorine (Cl) is a highly reactive chemical element and a member of the halogen group on the periodic table. In its pure, uncombined form, it exists as a diatomic molecule (Cl2)–a gas at room temperature with a distinct pungent odor. Due to its reactivity, it is rarely found in nature as a free element. While chlorine is an element that readily forms salts, it is not a salt itself.

In a chlorine pool, chlorine is added in a variety of chemical forms, which are often salts. When these compounds dissolve in water, they react to produce hypochlorous acid (HOCl) and hypochlorite ion (OCl–). These two compounds are the actual sanitizing agents that kill bacteria and other contaminants. Some ways of chlorinating pools:

- Sodium hypochlorite ($NaClO$): This is liquid chlorine, essentially a stronger version of household bleach. It is a salt.
- Calcium hypochlorite ($Ca(ClO)_2$): This is a solid, granular form of chlorine. It is also a salt.
- Trichloroisocyanuric acid ($C_3Cl_3N_3O_3$) and Dichloroisocyanuric acid ($C_3HCl_2N_3O_3$): These are the stabilized forms of chlorine found in tablets or sticks. They are not technically salts, but they contain chlorine and are often referred to as 'chlorine'.

Saltwater Pools

This is where the term 'salt' can be confusing. A saltwater pool is not chlorine-free. It uses a saltwater chlorinator (also called a salt cell) to produce its own chlorine. Salt is added to the pool water. The salt used is sodium chloride (NaCl), which is common table salt. This is a true salt.

In electrolysis, the saltwater passes through the chlorinator, which uses a process called electrolysis to split the sodium chloride molecules. Chlorine is created. This process generates chlorine gas (Cl2) and its dissolved forms, hypochlorous acid (HOCl) and sodium hypochlorite (NaClO). The same sanitizing agents found in traditional pools are created on-site.

So, while a saltwater pool uses salt, the chemical that actually cleans the water is chlorine, which is a byproduct of the electrolysis of that salt. The chlorine acts as a sanitizer. Hence, chlorine is the sanitizer. The chemical that kills germs and keeps the pool clean is hypochlorous acid (HOCl) and its related forms. Chlorine is often from a salt. The compounds used to add chlorine to a pool (like sodium hypochlorite or calcium hypochlorite) are salts. So saltwater pools still have chlorine. The salt in a saltwater pool is simply a way to generate chlorine automatically, so you don't have to add it manually.

Chlorinated Fresh Water Pool, Saltwater Pool, or Swimming in the Sea?

Each type of swimming environment–chlorinated freshwater pool, saltwater pool, and the sea–has its own set of health benefits and potential risks. The single 'healthiest' option depends on factors like the level of pollution in the other options, skin sensitivity, and pre-existing conditions.

When we consider chlorinated freshwater pools, we're looking at the historical champion of public sanitation. Its greatest advantage is undeniably its effective Sanitization: chlorine is a potent killer, a germicidal powerhouse capable of instantly neutralizing bacteria, viruses, and pathogens like E. coli and Salmonella, making crowded public environments safe. Because these pools operate in a controlled environment, water quality is vigilantly monitored and maintained, significantly lowering the risk of waterborne illness. Furthermore, chlorinated pools enjoy wide accessibility, offering a convenient and pervasive option for exercise and recreation across communities.

However, this chemical efficiency comes at a cost to the human body. The most common complaint is Skin, Eye, and Organ Irritation: chlorine and its irritating byproducts, known as chloramines, are notorious for stripping natural oils from skin and hair, resulting in dryness, redness, and itching. More concerningly, some research, like that by Xiao, Zhang, & Zhai, suggests that these byproducts may even cause harmful structural and functional changes in the delicate lining of the intestine, potentially affecting permeability.

Furthermore, these chloramines contribute to significant respiratory issues, particularly in poorly ventilated indoor spaces where the fumes irritate the lungs and airways. This is a special risk category for individuals with asthma, young children, and professionals like lifeguards who face chronic exposure. Lastly, the mere fact of chemical exposure prompts caution, as ongoing scientific investigation continues to explore potential links between long-term immersion in chlorinated water and various subtle health concerns.

The saltwater pool offers a sophisticated compromise, primarily championing a gentler experience. While a common misperception suggests they are chlorine-free, they are not; they simply generate chlorine on-site through electrolysis of salt. This method results in lower, more consistent concentrations of chlorine and far fewer irritating chloramines, making the water markedly gentler on skin and eyes. Many users, especially those with sensitive skin or conditions like eczema, find the water less harsh, and some report genuine therapeutic benefits—the mild salt level mimics a diluted Epsom salt bath, providing a soothing, softer feel that can aid muscle relaxation.

Crucially, because the chlorine is constantly generated and doesn't rely on adding volatile chemical shock treatments, saltwater pools feature a dramatically reduced chemical smell; the potent chlorine odor, which is actually the smell of irritating chloramines, is largely absent.

Yet, saltwater systems introduce new challenges. The most practical drawback is corrosion: the constant presence of salt, even at low concentrations, can accelerate the deterioration of expensive pool infrastructure, including metal ladders, handrails, fixtures, and certain types of decking. Moreover, saltwater pools are not chlorine-free, but simply a different way to chlorinate, meaning they share the same fundamental sanitizer, albeit at a lower concentration.

For those seeking the ultimate escape from man-made chemicals, swimming in the sea (or ocean) provides an unparalleled option. It is a completely natural and chemical-free environment, devoid of added synthetic treatments. The natural minerals present, particularly magnesium, have documented skin and respiratory health benefits; the saltwater is often soothing for the skin, and inhaling the salty air is an ancient remedy for respiratory ailments. Above all, sea swimming offers profound mental well-being benefits: the vast, natural setting is inherently calming, reducing stress and improving mood. The invigorating shock of cold water can even boost circulation and trigger a rush of feel-good endorphins, leading some enthusiasts to report a palpable electric charge after weeks or months of regular immersion.

Nevertheless, the sea trades chemical hazards for natural risks and uncertainty. The single greatest concern is the potential for contamination. Uncontrolled environments are vulnerable to agricultural runoff, untreated sewage, and industrial waste, increasing the risk of contracting gastrointestinal illnesses, ear infections, and skin rashes. Furthermore, the environment presents inherent environmental dangers, including strong currents, treacherous rip tides, and encounters with potentially harmful marine life. Finally, the temperature and conditions are uncontrolled; swimming comfort and safety are dictated solely by the weather and the season, making it an impractical or impossible option on many days of the year.

Unlike a pool, the sea contains sand, seaweed, and other natural debris that some people may find unpleasant. For pure sanitization and a controlled environment, a chlorinated freshwater pool is the most reliable choice, but it comes with the highest risk of skin, respiratory, and bodily irritation.

For a gentler, more comfortable swimming experience with fewer side effects, a saltwater pool could be the best pick. For mental and physical benefits from a natural, chemical-free environment, swimming in the sea is a great option. However, it's crucial to be mindful of water quality and safety, as it's an uncontrolled environment.

Relation of Cleaning Agents to Alzheimer's

The relationship between chlorinated pools and Alzheimer's disease is a complex and still-developing area of research. According to Xiao, Zhang, & Zhai, would not be a direct cause-and-effect, but rather a potential link through the byproducts of chlorine, which are also found in some household cleaning agents.

The role of disinfection byproducts (DBPs) is under scrutiny. When chlorine is added to water, it reacts with organic matter (like sweat, urine, and cosmetics) to create a class of chemicals called disinfection byproducts (DBPs). These include chloramines and trihalomethanes (THMs). These are the same kinds of chemical compounds found in many household cleaning and sanitizing agents. The relationship between chlorinated pools and Alzheimer's disease is a complex and still-developing area of research. According to Xiao, Zhang, & Zhai, the potential link to Alzheimer's disease is not from the chlorine itself, but from these DBPs.

Some DBPs have been identified as neurotoxic, meaning they can be harmful to the nervous system. The research explored how these compounds might affect the brain and contribute to neurodegenerative conditions. One study found that some DBPs, specifically halogenated acetic acids

(HAAs), can inhibit an enzyme critical for cellular energy production in neurons, which may lead to oxidative stress and cellular damage–processes that are thought to be involved in the development of Alzheimer's.

Oxidative stress and inflammation is another side effect. The brain is highly susceptible to oxidative stress, which occurs when there's an imbalance between free radicals and the body's ability to counteract them. Chronic inflammation and oxidative stress are known to play a significant role in the progression of Alzheimer's disease. Some studies suggest that exposure to DBPs can increase both of these, potentially contributing to the onset or acceleration of the disease.

Another area of concern is the blood-brain barrier (BBB), which is a protective layer of cells that prevents harmful substances from entering the brain. Some environmental toxins and pollutants, including certain DBPs, have been shown to disrupt the BBB, making the brain more vulnerable to damage from other sources.

There's also a broader 'hygiene hypothesis' being explored in relation to Alzheimer's. This theory suggests that as societies become more sanitized and have less exposure to a diverse range of microbes in the environment, the immune system may not develop properly. This could lead to an increased risk of autoimmune-like conditions and chronic inflammation, which are factors in many neurodegenerative diseases. While this is a broad hypothesis that doesn't focus solely on chlorine, the widespread use of disinfectants and sanitizers (like those found in pools and cleaning products) is a component of this cleaner environment.

Most of the research in this area shows a correlation. Although not a direct causation, a link has been observed, suggesting that exposure to chlorinated water or 'cleaning agents' causes Alzheimer's.

The concern is generally about low-level, chronic exposure over many years, not a single swim in a pool. While the public health benefits of chlorine disinfection are immense—it has virtually eliminated waterborne diseases like cholera and typhoid in many parts of the world—any potential risks from DBPs must be weighed against these proven benefits.

This field is still in its early stages. Scientists are actively working to better understand the mechanisms by which these chemicals might affect neurological health. With the relationship between chlorine pools and Alzheimer's being a topic of ongoing scientific study, the focus is on the long-term effects of disinfection byproducts (DBPs), which are also present in many cleaning agents. These byproducts have been linked to potential neurotoxicity and inflammation, which are known factors in neurodegenerative diseases like Alzheimer's.

The immediate financial sting of advanced technology versus the slow, steady bleed of chemical dependence. When directly comparing initial costs, the winner is unequivocally the traditional chlorine system. This classic method requires the absolute minimum upfront investment–perhaps just a simple chlorinator unit or, in the most basic scenario, nothing more than the chemicals themselves. It represents the low barrier to entry that has dominated the industry for decades. However, the advanced, chemical-reducing systems demand a far higher entry price. Standing above chlorine, the copper/silver ionizer requires a substantial initial outlay for the control unit and the specialized metallic electrodes. Yet, this cost pales in comparison to the investment needed for an ozone generator (Ozonator). Ozonators are sophisticated, high-performance units that necessitate professional installation and careful sizing to the pool's volume. The necessary hardware and labor combine to make the ozone system the undisputed claimant to the highest upfront cost among the three.

Despite the initial investment hurdle, the enduring power of these advanced systems lies in their dramatic, long-term assault on recurring chemical costs. A traditional chlorine pool is locked into a relationship of constant expenditure, perpetually buying chlorine, shock treatments, and algaecides to maintain water quality. This is where the copper filtration/ionization system fundamentally changes the equation, offering a clear and potent pathway to fiscal liberation. By releasing copper ions into the water, the system provides a continuous, powerful sanitizer and algaecide that allows a pool owner to slash their reliance on the primary sanitizer. Owners consistently see an 80% to 90% reduction in chlorine consumption, transforming the pool from a chemical sponge into a self-sustaining environment.

The main maintenance cost shifts from endless chemical shopping to the infrequent, necessary replacement of the copper electrodes, which wear down as they release ions. The savings generated by avoiding bulk chemical purchases often eclipse this replacement cost, meaning the initial price of the ionizer acts as an investment that slowly but surely pays for itself over a period of just a few years.

The ozone generator offers a similarly powerful path to long-term savings, but through the elemental force of oxidation. Utilizing an ozone cell, the system generates O3 molecules that are such powerful, instant oxidizers that they destroy contaminants upon contact, drastically diminishing the need for sanitizing chemicals. Like the copper system, the ozonator can reduce necessary chlorine inputs by a substantial margin, often 60% to 90%, generating massive annual savings. While the copper system's ongoing cost is electrode replacement, the ozonator's running cost is primarily low-level electrical consumption. Its main maintenance expense is the periodic replacement of the ozone generating cell or lamp, which is a significant, but infrequent, cost. Therefore, both the copper and the ozone systems offer a strong economic argument against the high-expenditure, high-maintenance routine of a pure chlorine pool. They transform the initial high cost from an obstacle into a calculated investment, positioning both advanced technologies as the demonstrably more economical choice in the long run when weighed against the endless, repetitive costs of chemical dependency.

20

Salt in Today's Western Medicine

"You are the salt of the earth."
– Matthew 5:13 *The Bible*

Salt's use in medicine has been largely replaced by pharmaceuticals, but its role has shifted rather than disappeared entirely. For millennia, salt was a cornerstone of medical practice, primarily due to its antiseptic and dehydrating properties. The development of modern pharmacology and a deeper understanding of human biology led to the creation of more targeted drugs.

Salt's Historical Role in Medicine

For most of human history, medical knowledge was based on observation and tradition. Salt, being a ubiquitous and powerful substance, was a natural choice for a wide range of treatments. Its medicinal applications were rooted in its ability to prevent infection.

Salt's hypertonic nature was used to draw water out of cells through osmosis. When applied to wounds, this dehydration effect killed or inhibited the growth of bacteria, making it a primitive but effective antiseptic. Ancient Egyptians, Greeks, and Romans all used salt to clean and treat wounds.

Inhaling steam from saltwater has been a common practice to help clear congestion and soothe sore throats and treat respiratory issues. This form of halotherapy is still used today in modern spas and clinics as a complementary therapy.

Salt has been used as an emetic to induce vomiting and as a laxative to treat constipation and aids digestion. Ancient medical texts often included salt in prescriptions for digestive ailments.

The Rise of Modern Pharmaceuticals

The shift from salt to pharmaceuticals began in the 19th and 20th centuries with the advent of scientific medicine. Instead of relying on general-purpose substances, scientists began to isolate and synthesize specific chemical compounds to target diseases with precision.

Salt's usage decreased with the discovery of bacteria and antibiotics. The germ theory of disease in the late 19th century revealed that infections were caused by specific microorganisms. This led to the development of powerful antibiotics, which could kill bacteria without causing the tissue irritation that concentrated salt solutions often did.

On the other hand, the creation of targeted drugs for issues like high blood pressure, which was once treated by a restrictive, low-salt diet, increased the use of table salt. Modern pharmaceuticals like diuretics and beta-blockers were developed; these drugs target the specific physiological mechanisms that cause hypertension, purporting a more effective and less disruptive treatment than broad dietary changes.

As medicine evolved, the human body's physiological mechanisms were understood at a cellular and molecular level. Conditions like goiter, once vaguely treated with salt, were found to be caused by iodine deficiency. This led to the creation of iodine supplements and the practice of iodizing salt, showing a shift from using salt as the primary medicine to using it as a vehicle for a more specific pharmaceutical.

Salt's Modern Medical Role

Today, salt is no longer a primary medical treatment for most ailments, but it is an essential component of modern medicine, particularly as a vehicle for drug delivery and for intravenous solutions.

A sterile solution of salt in water, normal saline solution is a cornerstone of hospital care. It is used for intravenous drips to rehydrate patients, to flush wounds, and as a carrier for administering medications directly into the bloodstream.

A significant number of modern drugs are manufactured as pharmaceutical salts. Converting an active drug ingredient into a salt form can improve its stability, solubility, and absorption rate, making it more effective. For example, some blood pressure medications are prescribed as a 'tartrate salt' to improve their effectiveness. Around 50% of modern drugs are formulated and sold as salts. This reliance on salts is not for their flavor but for their ability to significantly improve a drug's properties.

The Purpose of Drug Salts

The majority of drugs are either weak acids or weak bases. Their free form often has poor characteristics that make them difficult to administer effectively, for example, the drug may not dissolve sufficiently in water. This low solubility can prevent it from being absorbed into the bloodstream. Converting a drug into a salt can dramatically increase its solubility. In the case of ibuprofen, sodium ibuprofen is a salt form of ibuprofen that dissolves more easily and is absorbed faster than the free acid form.

Some drug molecules are unstable and can degrade over time, reducing their shelf life. Salt formation can stabilize the compound, ensuring medication with poor stability remains effective.

Reduced bioavailability is another problem that salt can solve. If a drug isn't soluble or stable, it can't be absorbed effectively by the body. This reduces its bioavailability, meaning a smaller amount of the drug reaches its intended target. The use of a salt form ensures that a consistent, effective dose is delivered. The most common drug salts are hydrochlorides, sodium salts, and sulfates. These are added to the active pharmaceutical ingredient (API) to enhance the drug's performance without altering its therapeutic action.

Other Medical Uses of Salts

Beyond drug formulation, salts are a foundational component of modern medicine. Normal saline, a 0.9% salt solution, is a vital medical fluid used for intravenous hydration, wound irrigation, and as a carrier for administering other drugs. It is considered an essential medicine by the World Health Organization.

Salts like potassium chloride and sodium bicarbonate are crucial for treating electrolyte imbalances, which can be caused by conditions like dehydration or kidney disease. These are administered for electrolyte replenishment to restore the body's normal functions.

Various salt-based solutions are used in surgical diagnostic procedures and as part of surgical processes to prevent infection and maintain tissue viability.

Betadine, a brand name for the antiseptic povidone-iodine (PVP-I), is widely used in hospitals for its broad-spectrum antimicrobial properties. It is effective against bacteria, viruses, fungi, and protozoa, making it a cornerstone of infection control.

In hospital settings, Betadine is used for pre-operative skin preparation to prepare a patient's skin before surgery. A surgical team will apply the solution to the patient's skin at the incision site to kill any microorganisms and reduce the risk of surgical site infection.

Betadine is used to clean and disinfect minor cuts, abrasions, and burns. It helps prevent infection in open wounds and promotes a clean environment for healing. It's also used as a medical professional hand scrub. While less common than alcohol-based rubs today, it is still used as a surgical scrub for healthcare professionals before procedures to decontaminate their hands. Betadine is also used to disinfect mucous membranes, such as in the mouth or vagina, before medical procedures.

While Betadine is a well-known antiseptic on both continents, its regulatory status and common usage differ. In the United States, Betadine (povidone-iodine) is a primary choice for pre-operative skin preparation. The Food and Drug Administration approves its use for a wide range of antiseptic applications. Many hospitals and surgical centers rely on it as a standard component of their infection control protocols, often in combination with alcohol. It's viewed as a reliable and effective product with a long history of successful use.

In the European Union, the use of povidone-iodine is still common, but there is a growing trend toward alternatives, particularly chlorhexidine-based solutions. This shift is driven by regulatory scrutiny. The European Medicines Agency has stricter regulations on antiseptic agents, and some countries have raised concerns about potential side effects like skin irritation and the possibility of iodine being absorbed into the body.

Some EU clinical guidelines now recommend chlorhexidine, a synthetic compound, as the preferred choice for pre-operative skin disinfection, as studies have shown it can provide longer-lasting antimicrobial activity on the skin. While rare, some patients can have allergic reactions to povidone-iodine, which has led some EU institutions to favor chlorhexidine as an alternative.

In summary, while Betadine is a staple in US hospitals for pre-operative disinfection, EU hospitals have shown a greater willingness to adopt and sometimes prefer chlorhexidine as a first-line alternative, although Betadine remains a common and viable option.

Salt is being re-evaluated for its therapeutic and functional properties in specialized, modern applications, often alongside or as a complement to existing technology.

Salt in Medicine

The use of salt in medicine is not a return to it as a primary treatment for disease. Instead, modern medicine is using it in new and more targeted ways. In halotherapy, doctors prescribe the practice of breathing in microscopic salt particles, often in a 'salt room'. It's not a cure but is used as a complementary therapy to help with respiratory conditions like asthma and allergies by reducing inflammation in the airways.

Salt is the foundation of wound healing. While antibiotics are the gold standard for treating infections, modern wound care often incorporates saline solutions or specialty salt dressings to manage certain types of chronic wounds. Saline solutions are used for irrigation, while salt dressings can help absorb excess fluid and promote a clean wound environment. Intravenous therapies also rely on salt.

Far from being replaced, saline solution (0.9% NaCl) remains the indispensable cornerstone of hospital care for rehydration and drug delivery. It's a fundamental part of emergency medicine, surgery, and routine patient care.

Oral Rehydration Therapy (ORT)

Oral rehydration solution (ORS) is a different mixture from a medical saline solution. An ORS is specifically designed for treating dehydration from diarrhea or vomiting. It contains a precise balance of salts and sugars. The sugar (glucose) in the solution helps your intestines absorb the water and sodium together, preventing the negative osmotic effect of salt alone. This makes ORS a safe and effective way to rehydrate, but it should not be confused with a basic saltwater solution.

The World Health Organization (WHO) and UNICEF recommend an Oral rehydration solution (ORS) with a particular concentration of salt, designed to effectively treat dehydration from diarrhea. ORS using common kitchen measurements contains 1 level teaspoon of salt dissolved into into 1 liter of clean water. The WHO rehydration solution actually recommended adding eight times as much glucose as salt, although a notable study recommended an alternative to glucose in oral rehydration solutions: Research by O. Fontaine, S.M. Gore, and N.F. Pierce, pointed to at least one solution significantly more effective than the standard WHO glucose solution in combating diarrhea.

The WHO warns that incorrect balance could result in a hypertonic solution is saltier than your blood, and that drinking this would cause water to be drawn out of your cells through osmosis, making dehydration worse, adding that it could also put a strain on kidneys as they try to excrete the excess salt.

Saline solution with only 0.9% NaCl remains the indispensable cornerstone of hospital care for rehydration and drug delivery. Normal saline is used for intravenous (IV) rehydration in clinical settings. It contains only sodium and chloride, without glucose or other electrolytes like potassium. The intravenous rehydration process is different from oral; the solution is infused directly into the bloodstream, bypassing the digestive system entirely. It is used when a patient cannot take fluids orally, such as due to vomiting, or when severe dehydration requires rapid intervention.

The Renin-Angiotensin-Aldosterone System

Research, including that of Dr. James DiNicolantonio, indicates that when your sodium levels drop too low, your body activates the renin-angiotensin-aldosterone system (RAS), a powerful hormonal response to pull sodium back into your blood. This system releases hormones like aldosterone and angiotensin II, which directly target fat cells. These hormones can cause fat cells to grow larger, become inflamed, and increase insulin resistance. He says they promote not only the growth of existing fat cells but also the differentiation of new ones.

While these hormones contribute to fat cell growth, their effect is minor compared to that of elevated insulin. A diet high in refined starches and sugars, combined with salt restriction, can significantly promote fat cell growth.

The Importance of Salt in the Body

Salt, which is about 40% sodium, is crucial for numerous bodily functions. It helps maintain fluid balance, and it's essential for nerve signaling, muscle movement, nutrient absorption in the gut, and cellular energy.

How to Use Salt for a Healthy Metabolism

According to Dr. Ben Bikman, to make salt work for you, you need to stop fearing it. The key is to consume the right kind of salt, not the highly processed table salt found in chips and packaged snacks. Instead, opt for natural salts like iodized sea salt, Himalayan pink salt, or Celtic salt, which contain beneficial trace minerals.

Bikman's simple trick is to dissolve 1/4 teaspoon of natural salt in a glass of warm water and drink it first thing in the morning. This provides your body with the sodium it needs without triggering fat-storing hormones. You can also add natural salt to your home-cooked meals. When your body has enough sodium, it stops overproducing the fat-promoting hormones aldosterone and angiotensin II, according to Bikman.

Salt and Blood Pressure

The link between salt and high blood pressure is a point of contention in the medical community. According to Dr. Bikman, for decades, salt has been unfairly blamed for high blood pressure. He says that for many, especially those who are insulin sensitive, cutting back on salt doesn't significantly lower blood pressure and may have unintended negative metabolic effects (like increasing insulin resistance). While salt consumption can temporarily cause water retention and a slight increase in blood pressure, multiple large studies have shown that for most people, cutting back on salt has a negligible effect on blood pressure, his research shows. The real culprit is often insulin resistance, which forces the body to hold onto salt and constricts blood vessels. He says, instead of cutting back on salt, the focus should be on reducing refined starches and sugars that spike insulin.

On the other hand, many large, well-respected health organizations (like the American Heart Association and the World Health Organization) still recommend reducing salt/sodium intake because moderate reduction has been shown in large studies and meta-analyses to significantly lower blood pressure in the majority of adults (both hypertensive and normotensive). The difference lies in the concept of salt sensitivity. Some people (including many with existing hypertension, kidney issues, or diabetes) are 'salt-sensitive', meaning their blood pressure reacts strongly to salt intake. Others are 'salt-resistant', and their blood pressure changes little with salt intake. The key issue is that it's difficult to know which category a person falls into without specialized testing.

The Risks of Quick Fixes

Many people turn to quick fixes for weight loss, such as drugs that work by blocking the hunger hormone GLP-1. However, these drugs can have significant side effects. A concerning amount of the weight lost is often from fat-free mass, including muscle and bone. The brain itself requires a certain amount of fat. There's also an increased risk of poor mental health outcomes, including a significant rise in the risk of depression and suicidal behavior. The body's innate systems already have natural tools to manage weight, according to

Dr. Bikman, and a safer, simpler approach is to support the body's natural metabolism, including with essential minerals like salt.

21

Black or Colored, Salt in Food

"Life is like cooking; too much salt is
bitter, too little is bland,
only when it's just right is it delicious."
"人生如烹飪，鹽多則苦，鹽少則淡，
恰到好處方為美味。"
Pinyin: "Rénshēng rú pēngrèn, yán duō
zé kǔ, yán shǎo zé dàn, qià dào hǎo chù
fāng wèi měiwèi." —Anon

Salt is the very essence of flavor. It is what transforms the bland into the savory, the flat into the complex. A dish without salt is a story without a soul. It sharpens, it brightens, it completes. It is the final touch of the chef's art and the secret weapon of the home cook.

The return to salt in food isn't about preservation but about flavor and artisanal quality. There is a growing trend of using traditional salt curing and fermentation methods in specialty foods. This is driven by a consumer desire for unique flavors, textures, and what's perceived as a more natural or 'authentic' product, rather than a necessity to prevent spoilage.

Artisan butchers and chefs are reviving traditional salt-curing techniques for charcuterie and cured meats like prosciutto, salami, and ham. This process imparts a complex flavor profile that refrigeration alone cannot. Pickled and Fermented vegetables, like kimchi and sauerkraut, and traditional salted pickles rely on salt to create an environment where beneficial bacteria can thrive, producing a unique sour flavor and texture.

Salt tenderizes meat through a process that involves both moisture and protein structure. Here's a breakdown of how it works:

Osmosis and Moisture Retention

When you apply salt to meat (either by dry brining or wet brining), it draws moisture out of the muscle fibers. The salt then dissolves in this liquid, creating a concentrated brine. The meat then reabsorbs this saltwater solution. This process increases the meat's overall moisture content before it's even cooked. When the meat is cooked, it will still lose some moisture, but because it started with a higher water content, the final result is a juicier, more succulent piece of meat.

Salt directly interacts with the proteins in the meat, specifically the muscle fibers and collagen (the connective tissue). This interaction causes the proteins to unwind, or 'denature'. This loosening of the protein structure prevents the muscle fibers from contracting as much during cooking. Since less contraction occurs, the meat remains more tender and less chewy.

Salt tenderizes meat by increasing juiciness. It helps the meat retain more moisture during the cooking process. Breaking down proteins: It alters the meat's protein structure, making it less tough and fibrous. This is why brining (either wet or dry) is a popular technique for cooking everything from chicken and pork to large cuts of beef like roasts.

Gourmet Salt

Now salt itself has become a gourmet item. Mark Kurlansky remarks in his book *Salt, A World History:*

> Gray salts, black salts, salts with any visible impurities are sought out and marketed for their colors, even though the tint usually means the presence of dirt.... Now there is talk of adding

> fluoride to salt for its health benefits. But modern people have seen too many chemicals and are ready to go back to eating dirt.

"Black Salt" refers to Kala Namak, a specific type of volcanic rock salt from the Himalayan regions (primarily India, Pakistan, and Nepal). This salt is black/pinkish-grey and has a distinct sulfurous, egg-like flavor due to its high sulfur and iron content, achieved through a unique firing process. It is used in vegan and South Asian cuisine.

According to an article in *Smithsonian Magazine* by A. Pester, entitled, "A restaurant in Japan is serving a $110 tasting menu featuring dirt,"

> The secret behind the dirt's lack of dirtiness may be its origins. The dirt comes from a company called Protoleaf, which applies its motto 'good grow green' to cocopeat made from coffee grinds, palm fiber and coconut shells and imported from India and Sri Lanka.

Geophagy is the deliberate and intentional practice of consuming earth or soil-like substances such as clay, chalk, or termite mounds. This practice, while appearing unusual to modern Western sensibilities, is a centuries-old behavior observed globally. A crucial distinction must be made between geophagy and the broader eating disorder known as pica. Pica is defined as the craving and consumption of non-food items, including dirt, hair, or chalk. While geophagy is a form of pica, it is often a socially or culturally appropriate practice, which complicates its simple classification as a pathological disorder.

In many communities, the consumption of earth is not considered a mental or physical health issue but rather a normative part of life, often with deeply ingrained traditional or medicinal justifications. This differentiation is fundamental to a nuanced analysis, as it moves the discussion from a purely medical one to a broader socio-cultural and anthropological examination.

Let Them Eat Dirt?

The practice of geophagy is deeply rooted in both human history and the natural world, suggesting a fundamental biological or behavioral drive. The oldest evidence of human geophagy dates back to a prehistoric site at Kalambo Falls, where a calcium-rich clay was found alongside the bones of Homo habilis. Ancient Greek physicians like Hippocrates and Pliny the Elder recorded the practice, and it has been documented among Native American tribes and in parts of Africa and the Southern United States for centuries. In Ghana, for instance, processed geophagic materials are given different names in various dialects, and they are ingested by pregnant women to suppress appetite and for their perceived health benefits. In Haiti, poor individuals consume 'bonbon tè', or mud cookies, as a survival mechanism. The tradition also spread to the Southern U.S.

with the institution of enslavement, where a 1942 survey in Mississippi found at least 25% of schoolchildren habitually ate earth.

This behavior is not unique to humans. Geophagy is a widespread adaptation in the animal kingdom, documented in over 100 primate species. Non-human primates use it for mineral supplementation, protection from parasites, and to help metabolize toxic compounds from leaves. Elephants, for example, consume soil from specific geological formations or termite mounds to acquire essential minerals such as sodium, potassium, and calcium. Similarly, mountain gorillas in Rwanda ingest clay during the dry season to help with the toxic compounds in the available vegetation. This broad evolutionary context underscores that consuming earth is not an isolated or bizarre human quirk but a deeply ingrained, cross-species behavior often driven by perceived physiological needs.

The Geophagy Paradox: Debated Benefits and Risks

For centuries, the practice of geophagy has been underpinned by two primary theoretical justifications: the 'Supplementation Hypothesis' and the 'Protection Hypothesis'. The supplementation theory posits that ingesting soil, which is often mineral-rich, provides essential nutrients that are otherwise lacking in a person's diet. This belief is prevalent in many cultures, particularly among pregnant women and children in parts of Africa, who crave clay for minerals such as phosphorus, potassium, magnesium, copper, zinc, manganese, and iron. In Haiti, mud cookies are traditionally consumed as an antacid and a source of calcium, especially by pregnant women.

The protection theory suggests that geophagy is a behavioral adaptation to mitigate health threats. Proponents of this view argue that certain clays can act as detoxifiers, providing a mud mask for our gut. This is because clay minerals have a high cation-exchange capacity, allowing them to bind to positively charged toxins in the acidic stomach, thereby rendering them safe for the body.

The clay mineral kaolinite, commonly found in geophagic materials, has been used in medications for diarrheal and intestinal problems. This protective function is also observed in animals; mountain gorillas, for instance, consume clay to help metabolize toxic compounds from certain plants, and ring-tailed lemurs use it for parasite control.

A Critical Analysis of Mineral Bioavailability

The widespread belief in geophagy as a nutritional supplement, a critical examination of scientific studies reveals a profound and counterintuitive paradox. Soils are rich in mineral nutrients including iron (Fe), Calcium (Ca), copper (Cu), and zinc (Zn).

Beyond the paradox of nutrient bioavailability, the practice of geophagy carries public health risks. The primary danger is the potential for heavy metal toxicity. Geophagic materials can contain toxic elements like lead (Pb), chromium (Cr), arsenic (As), cadmium (Cd), and mercury (Hg). These contaminants can occur naturally or result from industrial pollution. Lead in certain regions is a highly toxic metal that can cause extensive health problems, while inorganic arsenic is a lethal and carcinogenic protoplastic poison. Finally, there are significant physical risks. Consuming non-digestible material can lead to kidney complications.

The risk of parasitic and bacterial infections can be dealt with by baking the dirt at 200C for two hours. Unprepared, raw soil poses a much higher risk than traditionally processed earth. However, a specific study on geophagy among pregnant women in Zambia presented a fascinating nuance. While geophagy has been generally linked to helminthic diseases, this study found that the traditional practice of drying, roasting, or smoking the soil rendered it unsuitable for the survival of parasitic ova. This indicates that the risk of parasitic infection is not inherent to geophagy itself but is largely dependent on the preparation of the material.

The Modern Soil Movement: From Street Food to Star-Studded Cuisine

The dirt craze in high-end cuisine has seen chefs incorporate mineral rich, 'edible soil' or 'dirt' into their dishes. Another modern trend involves the literal consumption of real clays and chalks. This trend sees individuals purchasing and consuming unregulated products marketed as 'edible dirt'. The sellers promoting this dirt make a wide array of health claims, from improving gut health and clearer skin to better digestion, mood enhancement, and anti-aging properties.

But sometimes, geophagy is a necessity as in

the eating of Haitian 'Bonbon Tè'. The Haitian 'bonbon tè', or mud cookie, serves as an interesting case study for understanding geophagy in the context of extreme necessity. Made from a simple mixture of edible clay from the Central Plateau, water, salt, and vegetable shortening, these sun-dried discs have become a regular meal for Haitians facing food shortages and extreme poverty.

The existence of bonbon tè exemplifies a typology of geophagy based on motivation. While traditional geophagy is often linked to cultural beliefs or perceived medicinal benefits, and modern 'dirt' cuisine is an aesthetic pursuit, the Haitian mud cookie is a geophagy of sheer survival. It is a stark reminder that for some of the world's most vulnerable populations, 'eating dirt' is a last resort to stave off hunger and fill the stomach, even with the known risks of malnutrition, stomach pain, and dental issues.

Inquiry is warranted to investigate the specific processing methods used in traditional geophagic practices (e.g., drying, roasting) to better understand their potential for mitigating health risks, particularly in relation to parasitic ova. This research could offer a pathway for culturally appropriate harm reduction strategies.

A craving for mineral salts might lead one to eat dirt, while others spend a fortune to have their own salts perfectly processed. At least these cookies contain a full daily dose of mineral salts, a fact many modern, highly-processed foods can't claim.

22

The Mysteries of Salt

"You are the salt of the earth.
But if the salt loses its saltiness,
how can it be made salty again? It is no
longer good for anything, except to be
thrown out and trampled underfoot."
— *The Bible*, Matthew 5:13

In this famous quote from the Sermon on the Mount, Jesus uses salt as a metaphor for the preserving and flavoring influence his disciples should have on the world, but also alludes to its use in war. A chilling myth of salting the earth–the ultimate act of annihilation–began before the time of the Romans, which may have happened after Rome destroyed Carthage in 146 BCE and cursed the Earth. It echoes a far older, darker ritual of destruction rooted in the ancient Near East. A malicious custom of salting the Earth appears to be less of a practical agricultural technique, and more of a profound symbolic curse meant to invoke desolation.

Its most notorious historical precedent occurs in *The Bible* the moment the judge Abimelech captured the rebellious city of Shechem. After leveling its walls and slaughtering its inhabitants, he performed the final, horrifying ceremony, sowing the very ground with salt–a permanent, ritualistic pronouncement that the site was cursed, incapable of rebirth, and forever consigned to barren ruin.

This was more likely a ritual, since applying enough salt to an entire agricultural region to render it barren would have required a massive, logistically difficult, and extraordinarily expensive effort, especially since salt was a valuable commodity at the time.

So the unresolved saga of salt unfolds, with its benefits cast into stark relief against shadowy facets of its consequences. Salt's duality can also be seen in the mysteries scientists are currently studying to unlock.

In NaCl, we encounter a substance of profound duality, offering benefits, but also destruction. The shadow side of salt for just as salt speaks of evaporated ancient seas, it is now speaking of our future–a karmic key unlocking the next epoch of science and survival. Scientists are currently investigating several major mysteries about salt (NaCl) across chemistry, biology, and planetary science, moving far beyond its simple role as a condiment.

In medicine, scientists seek the true sweet (or salt) spot of life, navigating the mystery of salt's double edge: how it can simultaneously sustain life by governing nervous function, yet turn traitor, triggering brain inflammation and autoimmune disease in the genetically sensitive. We are only just mapping the gut-brain-salt axis, tracing a path of chronic disease that begins with an imbalance in our microbial inner space.

Esoteric & Alchemical Mysteries

These are not mysteries in the sense of current conventional laboratory research, but rather properties historically attributed to salt that continue to inform spiritual and sometimes fringe scientific inquiry:

In historical alchemy, as per Paracelsus, salt was considered one of the three foundational substances along with Mercury/Spirit and Sulfur/Soul, in the alchemical 'body'. Salt represents the physical, stable 'body' or the material essence that survives the fire and prevents dissolution ('corruption and preservation against corruption'). The mystery here is the symbolic and philosophical parallel between salt's chemical properties and the integrity of the human body and soul.

Many traditions (including Catholicism, Buddhism, Shinto, and various folk magic systems) use salt for purification of energy fields, warding off evil, and creating protective boundaries. Modern interpretations of this practice often posit that salt absorbs or neutralizes 'negative energy' or helps maintain a clear and balanced human energy field. The scientific mystery would be if there is any measurable physical interaction–perhaps through its negative ionic charge or hygroscopic properties–that corresponds to these traditional effects.

Salt's 'salvation' symbolism and historical use as a non-perishable preservative led it to symbolize eternity, loyalty, and an enduring covenant (as seen in the Biblical 'covenant of salt'). The enigma is how a common mineral came to hold such universal, deep metaphysical significance across disparate global cultures, suggesting an inherent quality recognized by humanity.

Salt's geometric, crystalline structure is stable and can be seen as an expression of order. In esoteric thought, crystals are often associated with energy storage and amplification. While unproven by mainstream science, the possibility of salt's crystal lattice resonating and interacting with subtle energies remains a topic in certain modern 'frequency medicine' or vibrational healing practices.

The Salt Spot

Health and biological mysteries abound as a significant portion of current research focuses on the complex and sometimes contradictory role of sodium and chloride ions in the human body, specifically obsessing on salt sensitivity and blood pressure. Unraveling the exact mechanisms that cause some people to be 'salt-sensitive', meaning their blood pressure is significantly affected by salt intake, while others are 'salt-resistant'. This includes studying the role of genetics and various hormones.

Any way you slice it, optimal intake of salt is under investigation. Scientists are attempting to determine the true—but it would be an *insalt* to call it a 'sweet spot'—rather, a 'salt spot'—for regarding sodium consumption, as mentioned earlier, new data suggests that consuming too little salt may also be associated with negative health consequences, such as damage to the vascular system, whereas most research has traditionally focused on the pitfalls of too much salt (for example, a study by Strazzullo, P. et al. (2009), "Salt sensitivity of blood pressure: a systematic review and meta-analysis of controlled trials," on blood pressure and cardiovascular risks), finding excessive salt consumption may raise the risk of stomach cancer, which is one of the many health effects of salt currently under scientific scrutiny.

Another area of investigation is the brain-salt connection. Investigating new findings that suggest a high-salt diet can trigger brain inflammation, which in turn drives up blood pressure by affecting the release of certain hormones like vasopressin. This could point to the brain as a new target for treating certain types of hypertension.

Scientists are also exploring the link between high salt intake and autoimmune diseases. Researchers are trying to determine how salt influences the differentiation and function of critical immune cells (T cells), and the development or activity of immune-mediated diseases (IMDs), such as multiple sclerosis (MS) and Rheumatoid Arthritis (RA).

Geochemistry & Planetary Science Mysteries

Scientists are exploring the behavior of salt under extreme conditions to better understand celestial bodies and 'forbidden' salt structures. Studying the possibility of exotic forms of salt (Na3 Cl, NaCl3, etc.) that are 'forbidden' by classical chemistry rules but can be stable under the extreme pressures and low temperatures found inside planets and icy moons (like Jupiter's Europa and Ganymede).

Icy Moon Composition

Characterizing newly discovered 'hyperhydrated' forms of frozen saltwater (with a much higher proportion of water molecules than naturally found on Earth). This is aimed at solving the long-standing mystery of unidentified spectral signatures observed on the surfaces of icy moons, which suggests a unique, high-water-content salt structure.

Ancient Life Preservation

Interrogating ancient DNA and cellulose trapped within salt crystals in geologic deposits that are hundreds of millions of years old. Scientists are trying to determine how salt crystals provide such an exceptional environment for the preservation of organic material.

Research continues on fundamental chemical mysteries, such as the basic interaction of salt and water that are essential to all biological and chemical processes. Scientists are investigating solubility and how the ions in salt (sodium and chloride) affect the structure of water at a molecular level. While the sodium ion's effect is generally local, scientists are still trying to understand the farther-reaching impact of the chloride ion and whether these subtle changes are sufficient to explain the different solubility of complex biomolecules in salty water.

Misunderstood Salt

Unresolved mysteries of salt include its less understood interactions in physics, biology, and its historically attributed metaphysical properties. Advanced biological and neurocognitive puzzles abound. This area goes beyond blood pressure to examine salt's deepest influence on the body and mind.

Scientists are actively investigating how a high-salt diet can directly and independently impair cognitive function and memory in the absence of high blood pressure. Research suggests a mechanism where excessive sodium leads to dementia and cognitive decline by causing a deficiency of nitric oxide in the brain's blood vessels, which in turn destabilizes the tau protein—a key hallmark of Alzheimer's disease. A complex mystery related to the gut-brain-salt axis is the proposed link between high dietary salt, changes to the gut microbiome and immune response, and subsequent inflammation in the brain. Researchers are tracing how immune-related molecules released in the gut travel to the brain's blood vessels, restricting blood flow and leading to cognitive deficits.

Electrolyte balance and consciousness is another area of intense research. While well-established that sodium and potassium ions are crucial for electrical signal conduction in the nervous system, the precise way the body maintains this exquisite, delicate balance of electrolytes (salt) is still being mapped. Any major disturbance in this balance leads to altered mental status, hallucinations, and seizures, highlighting a profound, yet not fully understood, link between salt concentration and conscious function.

Beyond its cardiovascular effects, scientists are studying salt's role in influencing metabolism, energy balance, and thermogenesis (heat production) to see if high salt intake may lead to effects like leptin resistance (disrupting satiety signals) and altered hormonal factors, showing a deeper role in energy regulation.

Fuel and Energy Roles

The arrival of CATL's sodium-ion battery

has been officially confirmed. China's new technology is expected to create serious competition for existing Lithium Iron Phosphate (LFP) batteries, potentially marking a shift away from LFP. While the hype has fueled exaggerated claims–no, it's not 90% cheaper–the truth is no less dramatic: at the cell level, the new battery, dubbed NXTra, currently costs a remarkable ~$19 per kilowatt-hour. This ~65% drop from today's LFP cells, has the potential to fall further to just ~$10 per kilowatt-hour.

But the staggering price is only the beginning; the true revolution lies in the battery's performance and longevity. CATL claims the NXTra can deliver an astonishing 3.6 million miles (≈5.8 million km) of driving while retaining 85% capacity, outlasting even the most durable LFP packs by three to six times, meaning the car itself will wear out long before the power source. Furthermore, sodium has conquered its former Achilles' heel: energy density. At 175 Watt-hours per kilogram, the NXTra now surpasses even BYD's current blade battery, matching lithium-ion's usable range while being safer, cheaper, and far longer-lasting, with flawless operation from −40° to +70° C.

This is more than just an EV upgrade; it's a fundamental redefinition of the entire energy landscape, utilizing abundant materials like sodium, aluminum, and carbon to enable a future of genuine energy independence and local production. For renewable energy, this means coupling the world's cheapest power source, solar, with storage that lasts half a century and costs a fraction of the price, effectively rendering coal, gas, and even nuclear obsolete. CATL has also cleverly announced a hybrid system called frivoid, blending sodium and lithium-ion cells in the same pack to enable a smooth industry transition, but the ultimate scaling advantage is that the NXTra cells can be manufactured on the existing LFP production lines with minimal retooling. This ability to pivot current multi-billion dollar factories to the new chemistry almost immediately is how quickly the sodium-ion battery will come to market, ensuring that within the next few years, this breakthrough may be seen as the moment lithium's decades-long dominance began to fade.

Fuel and energy are areas of cutting-edge research. Beyond its life-sustaining properties, salt's crystalline integrity proves essential for the next age of energy. It is the unblinking, thermal body for advanced nuclear reactors, where molten salt absorbs the fierce fire of the atom, offering inherent safety. It is the vast, low-cost battery for solar power, a silent subterranean cistern holding the heat of the sun to be dispensed at midnight.

Salt is proving to be a critical material for next-generation, low-carbon energy technologies, such as advanced nuclear energy (Molten Salt Reactors - MSRs). Nuclear fuel/coolant melts and uses certain chloride (like NaCl) and fluoride salts (like LiF-BeF2, or FLiBe) as the primary coolant, or even as the medium in which the nuclear fuel itself is dissolved. This enhances safety. MSRs operate at low pressure and very high temperatures, which eliminates the risk of a high-pressure steam explosion (a risk in conventional reactors). The liquid fuel can be passively drained into a containment vessel, where it solidifies and quenches the reaction, offering inherent safety advantages.

Molten salts can facilitate pyroprocessing, a chemical recycling technique that separates usable fuel (actinides) from fission products, potentially closing the nuclear fuel cycle and significantly reducing the volume and long-term radiotoxicity of nuclear waste. Thus, salts enable fuel cycle flexibility.

Thermal Energy Storage (Molten Salt Batteries)

Salt is useful for renewable energy buffering.

In Concentrated Solar Power (CSP) plants, molten salt mixtures (typically a blend of sodium nitrate and potassium nitrate) are heated to 565 C to 600 C by focused sunlight. This hot salt is stored in insulated tanks, acting as a massive, low-cost 'thermal battery'. This stored heat can then be used to generate steam and electricity on demand, day or night, effectively solving the problem of solar intermittency. Molten salt is also being explored as a heat-transfer medium. Salt fluid is used to convert existing coal-fired power plants to use stored renewable heat, repurposing old infrastructure.

Subsurface Energy Storage & Carbon Management

Salt has shown promise in the area of hydrogen storage. Large underground salt deposits (salt domes and caverns) are geologically ideal, proven, and leak-tight containers for storing massive quantities of hydrogen gas (H_2), a key clean fuel. This is essential for building the infrastructure of the future hydrogen economy.

Salt formations have a high thermal conductivity. Researchers are studying how to leverage the heat-conducting properties of underground salt to efficiently channel heat from warmer underlying rocks to the surface, aiding in the adoption of next-generation geothermal energy technology.

Osmotic Energy (Salinity Gradient Power)

There is a direct energy application reliant on salt, called 'Blue Energy': Osmotic power harnesses the energy released when waters of different salt concentrations (salinity gradients) are mixed, typically at the confluence of freshwater rivers and saltwater seas.

A similar area of research is pressure retarded osmosis (PRO). Commercial applications are developing systems that utilize highly concentrated brine and low-salinity water to generate pressure that drives a turbine, creating CO2-free energy. Danish inventor Jørgen Mads Clausen founded SaltPower in 2015, based on his insight that the high salt concentration in geothermal water could make pressure retarded osmosis–a technique previously deemed unprofitable with seawater–economically viable. He initiated university research that confirmed the profitability of using highly saline water to generate renewable, non-fluctuating energy. Clausen remains the active owner as SaltPower scales up its operations, now partnering with industrial salt producers for its first commercial-scale installation.

Salt Solutions and Electrical Conductivity

Researchers are exploring methods to use saltwater to produce energy. The most significant way saltwater is utilized to produce fuel is splitting seawater into hydrogen gas

(H_2) and oxygen (O_2) through a process called electrolysis. Hydrogen is a clean fuel that can be used in fuel cells to generate electricity. But the salt in the water is more of a challenge than a help. Traditional electrolysis requires highly purified, deionized water because the chloride ions (Cl^-) in seawater are highly corrosive and quickly degrade the electrodes (anodes) in the electrolyzer.

The salt itself is more helpful in another example, the saltwater fuel cell (metal-air battery), where the entire saltwater solution (or seawater) acts as the electrolyte. In order to conduct electricity, chloride ions (Cl^-) in saltwater are used as the electrolyte in the saltwater fuel cell or metal-air battery, particularly those using a magnesium anode. In short, while the salt (NaCl) is added to make the water a strong electrolyte, the resulting Cl^- ions are the essential negative charge carriers within that electrolyte.

This system generates electricity via a clean, simple chemical reaction. The saltwater acts as an electrolyte, allowing a magnesium anode to oxidize (corrode) in the presence of oxygen from the air. This chemical reaction releases electrons, creating an electrical current. The primary role of Cl– (along with Na+) is to ensure ionic conductivity. During discharge, the magnesium anode releases Mg2+ ions, and the cathode produces OH– ions (from the reduction of oxygen). The Cl– ions must migrate through the water to balance the charge generated by the electrode reactions, completing the internal circuit and allowing the battery to function.

While these cells are currently used mostly for small-scale applications and educational toys (like the saltwater fuel cell robot), the technology is an example of harnessing the chemical potential of common materials for energy. The primary fuel source is the magnesium plate, which is consumed over time.

The chloride ion (Cl^-) in a saltwater fuel cell (specifically, a Magnesium-Air Battery or Mg-Air MAFC) plays a dual role: it is helpful for the battery's basic operation but is also a major nuisance that limits performance and durability.

In these examples, the salt is not the fuel, but rather a component of the electrolyte. In the saltwater fuel cell (metal-air battery), what serves as the fuel is magnesium; the oxidant is oxygen; and the electrolyte is the salt.

The fuel that is consumed to produce electricity is the metal anode (e.g., magnesium (Mg), but can this also be aluminum (Al), or zinc (Zn)). The metal is oxidized (sacrificed), releasing electrons into the external circuit. The oxidant is oxygen from the air or water. A reduction reaction takes place whereby oxygen consumes the electrons from the circuit.

The electrolyte is saltwater, as an ionic

conductor. Its function is to conduct ions (Na^+, Cl^-, Mg^{2+}, OH^-) between the anode and cathode inside the battery to maintain charge neutrality. It is not consumed as a fuel in the way the metal is. While the salt is absolutely necessary for the battery to work (since pure water is a poor conductor), it is best described as the ionic conductor or activator, not the fuel.

In the CATL sodium-ion battery, The positive sodium ions move back and forth between the battery's cathode and anode to release and store electrical energy. They are a crucial active material that enables the battery to function, but they are not burned up like fuel. Istead, sodium ions travel between the cathode and anode through a liquid electrolyte (a salt dissolved in a solvent) to charge and discharge. The sodium itself is part of the core battery chemistry, not a consumable 'fuel' for combustion.

23

Connecting the Dots

"He had been bruised so badly that the eyes of strangers lacerated him like salt."
— James Baldwin, *Giovanni's Room.*
This 1956 novel captures the power of salt.

Turning from the macroscopic world to the truly microscopic, while Clausen's company harnesses high salt concentration to generate megawatts of power, other researchers are focused on exploiting biological molecules for communication itself.

The field of molecular communication has emerged, specifically exemplified by the work of Farsad, Eckford, and Hatzinakos, who explore its potential in a paper entitled, "Molecular communication in nanonetworks: the potential of engineered bacteria." Their research in nanobiology explores how ions, including sodium and chloride, might participate in an 'ionic signaling' or 'molecular communication' system that exists alongside electrochemical signaling in the body, suggesting a more complex, physical level of internal communication.

Yet an even tinier connection, popular in quantum physics (as well as esoteric circles), is the theory of the zero-point energy (ZPE) connection that Max Plank and Einstein both worked on. Central to quantum physics theories, the fundamental crystalline structure and highly stable nature of salt (sodium chloride) has led to speculative connections to the idea of a stable point of energy or a resonant structure that could hypothetically interact with or mediate 'zero-point energy' or vacuum energy. The connection between the complex quantum physics concept of zero-point energy (ZPE) and the structure of salt was the basis of several of their experiments in the 1920s.

The historical connection between salt and the concept of zero-point energy in this history of physics is that a salt's conductive nature was used in an early, critical test to challenge the zero point concept.

The concept of zero-point energy—the theory that even at absolute zero, motion never stops and 'nothingness has energy'—was Max Planck's attempt at a mathematical feat. Around 1910, Planck sought to show how light was emitted in discrete amounts but absorbed continuously. This 'second quantum theory', though ultimately wrong, mathematically produced the same black-body radiation law as his first theory, but with an inescapable addition: a residual energy ($E=21h\nu$) for all oscillators, even at 0 K.

Planck initially suspected this extra term would have no experimental consequences, dismissing it as a mere mathematical blandishment.

Salt's Crystalline Challenge

The intervention of salt, and other elemental compounds, came at the 1911 Solvay Conference. While the theoretical debate raged over Planck's new theory, the experimentalists, viewing the issue as a practical art, raised immediate objections. In the spirit of Calvert's salt merchants, scientists set about attacking Planck's zero point theory that stillness was mere illusion. What they couldn't disprove mathematically, they attempted to disprove experimentally, using toxic super conductors. Heike Kamerlingh Onnes found that the existence of this residual zero-point energy would directly conflict with measurements of the electric conductivity of mercury.

The core of the argument lay in the fact that the zero-point energy predicted a minimum level of vibrational motion in all matter. In highly conductive materials, like mercury, which, when cooled, becomes a superconductor, the unimpeded flow of current—a pure, low-entropy state—would be fundamentally compromised by this residual quantum 'wobble'. In this sense, the perfect, concentrated conductivity of a cooled element—a kind of electrical masterpiece—was threatened by the constant motion of the zero-point energy, suggesting that the existence of movement was incompatible with the pristine act of perfect conduction.

While this specific experimental challenge did not immediately kill the zero point idea (it was later disproved by the rise of Bohr's symmetric atomic model), On April 8, 1911, at his laboratory in Leiden, Netherlands, Heike Kamerlingh Onnes (1853–1926) used a material with higher conductivity to argue against Planck's 'second quantum theory'.

Onnes challenged Planck's conductivity theory by using superconducting mercury—not table salt (NaCl) itself. Salt's toxic crystalline kin, the elemental metal mercury, is crystalline when frozen and cooled below its freezing point. Particularly in this superconducting state, mercury provided the first electrical test against Planck's new, unsettling quantum power. But Onnes proved wrong, and his prediction, that zero-point energy would destroy superconductivity, was ultimately disproved, making way for the development of modern quantum mechanics.

Further correction to Planck's theory disproved his asymmetric idea in 1913 with Niels Bohr's atomic model. Bohr's model explained that spectral lines are produced by electrons jumping between discrete energy levels:

- Emission occurs when an electron drops to a lower energy level, giving off a light quantum (photon).

- Absorption occurs when an electron jumps to a higher energy level, taking in a light quantum.

Crucially, these two processes are symmetric. A specific jump down (emission) is balanced by the possibility of the exact same jump up (absorption). This symmetric, fully quantum view proved Planck's asymmetric (discrete emission, continuous absorption) 'second quantum theory' to be incorrect, validating the thermodynamic requirement that emission and absorption must be perfectly balanced in equilibrium.

No one was able to mathematically prove the existence of zero-point energy, but it has been proven experimentally. The proof of zero-point energy (ZPE) is not attributed to a single person, but rather to a series of theoretical developments and landmark experimental observations that provided evidence for its effects. And yes, they were done using salt.

The most direct scientific experiments relating salt (NaCl) crystals, which are non-toxic to the human body, and zero-point energy were performed in the early days of quantum mechanics (1920s) in the zero-point energy in Rock-Salt Lattice experiment to validate zero-point energy's existence in solid matter. The experiment was done using an X-ray diffraction method on the NaCl (rock-salt) crystal lattice. Researchers, including James, Waller, and Hartree (1928), determined whether the atoms in the crystal lattice possessed kinetic energy even at absolute zero (0 K). They did this by measuring the scattering of X-rays by NaCl crystals at very low temperatures. Atomic vibration (even ZPE vibration) reduces the intensity of the scattered X-rays.

By extrapolating the scattering data to absolute zero, they could check whether the calculated scattering power matched the theoretical value that assumed the existence of zero-point energy in the lattice (i.e., that the atoms are still vibrating) or the classical value (where all motion ceases). The results supported the quantum mechanical prediction that zero-point energy exists and that the atoms never cease motion. This was a key piece of early evidence for zero-point energy in solids.

A part of Planck's original formula was proven scientifically later in 1924 using another super conductor, boron monoxide, which like mercury is toxic to the human body. Final validation came at last, proving zero-point energy to be an absolutely inescapable feature of modern quantum mechanics. Although proof arrived not just from one source, but from a relentless proliferation of phenomena, Robert S. Mulliken connected the dots to the necessary inclusion of Planck's E=21hν by precisely measuring the spectroscopy of molecules like (toxic) boron monoxide, and published it in "The Isotope Effect in Band Spectra, II: The Spectrum of Boron Monoxide" (published in

Physical Review, 1925).

Later, the dramatic, final validations came from the extreme: the Casimir effect, where two uncharged plates attract due to the energy of the vacuum, and the unique quantum behavior of liquid helium, which uses its zero-point energy to perform the retrograde feat of refusing to freeze, even at absolute zero pressure. The ultimate truth is that nothingness itself harbors energy. The Casimir effect, which was predicted by Dutch physicist Hendrik Casimir in 1948—if not ages earlier by the originators of the yin-yang symbol—was definitively proved and quantitatively measured to a high degree of accuracy nearly fifty years later, in 1997, by Steven K. Lamoreaux.

Thus, theoretical error proved to be a concentrated, enduring truth. zero-point energy is now an absolutely inescapable feature of modern quantum mechanics, a fundamental observation that validates the theory. Proof arrives not through a single, elegant experiment, but through the continuous, relentless proliferation of quantum phenomena: the Casimir effect, discovered in an experiment where two uncharged plates attracted each other due to the energy of the vacuum; and the stubborn inability of helium to freeze even at absolute zero pressure.

Plank's formula (E0=21hν) has been necessarily included in the spectroscopy of molecular systems. It is the ultimate equation, quietly satisfying the repeated result that nothingness itself harbors energy.

Planck's Actual Formula

The fundamental equation introduced by Max Planck in 1900, which started quantum theory, relates the energy of a quantum (a photon or a discrete unit of energy) to its frequency.

$E = h\nu$

- E is the Energy of the quantum (photon), in Joules (J).
- h is Planck's constant, a fundamental constant of nature ($h \approx 6.626 \times 10{-}34$ J·s).
- ν (the Greek letter nu) is the frequency of the radiation, in Hertz (Hz or s−1).

This equation states that energy is quantized—it can only exist in discrete packets (multiples of hν).

What is Zero-Point energy?

In classical physics, an oscillating system (like a pendulum or a spring) could theoretically have zero energy and be perfectly still at absolute zero temperature (0 K).

However, in quantum mechanics, the Heisenberg uncertainty principle forbids a particle from being perfectly still. If it were still (zero momentum), its position would be perfectly known, which violates the principle. Therefore, every quantum harmonic oscillator must retain a minimum amount of energy even in its lowest possible energy state (n=0, the ground state).

Table salt (sodium chloride) is known for its perfect, repeating, three-dimensional crystalline structure. It's a very stable, orderly compound. ZPE is the minimum energy that a quantum system (such as a particle) can possess. Even in the absolute vacuum of space, at a temperature of absolute zero, there's still a tiny, residual amount of energy —it's never truly 'zero'. Some theories exaggerate this to an enormous, untapped source of 'vacuum energy'.

This theory suggests that because salt is so perfectly ordered and stable, like a high-quality antenna or resonator, its crystalline structure could lock onto or resonate with this vast, underlying zero-point energy field.

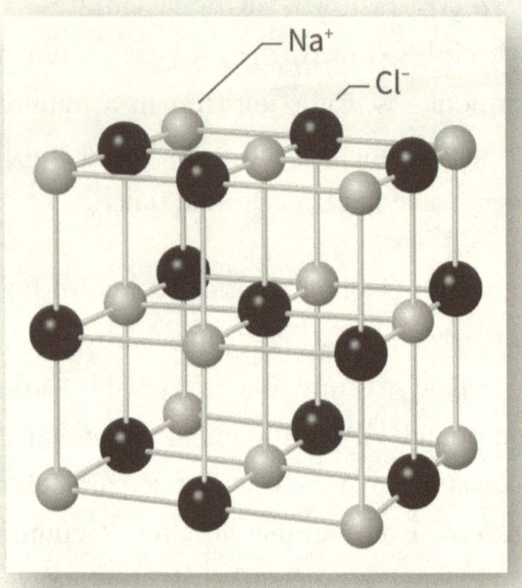

CRYSTALLINE STRUCTURE FOR HALITE
The mineral (natural) form of sodium chloride ($NaCl$)

Essentially, the stable, orderly structure of the salt is hypothesized to act as a mediator or gateway to tap into the untapped energy of seemingly 'empty' space.

Countless cultural and historical accounts bear testimony to the special powers of salts.

Salt's properties of preservation and purity have been linked to a universal, practical and intuitive understanding of its opposition to corruption and decay. The deep-seated and widespread mythological anathema and aversion to salt by 'evil' entities remains a peculiar cross-cultural psychological and historical mystery. Even in the depths of space, salt is a mystery whispered by the icy moons of Jupiter, where under crushing planetary pressures, salt forms structures 'forbidden' on Earth—exotic chemical geometries that only the forces of an extra-terrestrial gas giant could possibly reveal, suggesting a profound and stable purity might govern the composition of those jovial seas.

Further research must be done, not just by breaking salt down into its ionic components, but also on the applications of salt in its intact, lattice form. Cymatics, the study of visible sound and vibration (e.g., patterns formed by sound waves in substances such as sand, water, or salt, may sound out the relation of salt to the absorption of nutrients.

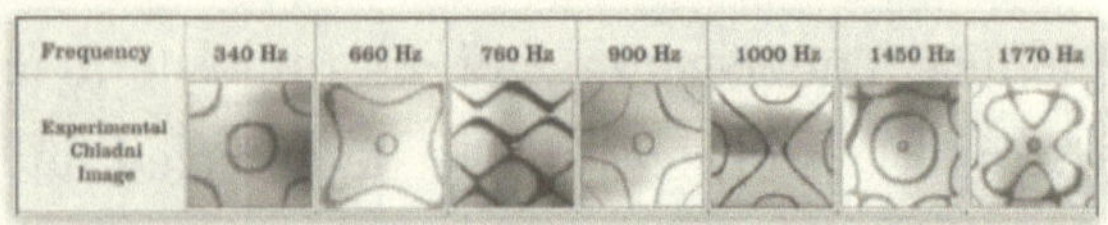

A modern demonstration of cymatics using frequencies to shape sand, from Wani, K., Saini, A. and Salunke, S. (2024) "Modal Frequency Prediction of Chladni Patterns Using Machine Learning," Journal of Modern Physics, 15(05), pp. 297–308

By ringing a Nepalese bowl before meals, we could be tapping into the theoretical zero-point energy (ZPE) connection. In some theories, the fundamental crystalline structure and highly stable nature of salt (sodium chloride) have led to speculative connections to the idea of a stable point of energy or a resonant structure that could hypothetically interact with or mediate zero-point energy, or vacuum energy. The connection between the complex quantum physics concept of zero-point energy (ZPE) and the structure of salt is a highly speculative idea from the fringes of science explored in the theories of scientists such as Puthoff.

Dots must be connected between concepts like cymatics, bio-absorption, Nepalese bowls, crystalline structure, salt, and zero-point energy (ZPE)—beyond the frontiers of peer-reviewed scientific evidence to establish the relationship between salt, cymatics, and ZPE. If we can apply Calvert's analysis of the human nature of salt merchants ("...the trade, instead of being ruled by common sense and business experience...") to that of the industrialists funding scientific 'research-and-development', we can see that there's plenty of room for research and learning, when it comes to salts. Hence, the proposed relationship links a mechanical vibration (the bowl's sound) to a change in matter (nutrient absorption) via the theoretical, non-material ZPE field, with salt acting as the hypothetical resonant intermediary.

It is well known that electrolytes, such as

sodium (Na+), are crucial for conveying other nutrients like glucose and amino acids into the body's cells, mainly by powering a process called secondary active transport. This mechanism starts with the Na+ /K+ - ATPase pump, which uses energy to create a steep electrochemical gradient by forcing a high concentration of Na+ outside the cell. The potential energy stored in this gradient is then released as Na+ rushes back into the cell through a cotransporter protein, and this energy is simultaneously harnessed to 'drag' the desired nutrient (e.g., a glucose molecule) inward, effectively moving the nutrient against its own concentration gradient.

Cymatics could act as a similar catalyst. Just as a dinner bell might be struck to call people to a meal, using a mallet on a singing bowl to signal the start of a family or communal meal has become a modern, New Age practice that draws on Himalayan and Tibetan cultures' use of the singing bowl for a moment of mindfulness/blessing. The logical hypothesis would be that the vibration resonating with the salt in the meal conveys enhanced nutrient absorption as the outcome.

Cymatics is hypothesized to interact with the fundamental crystalline structure of salt (NaCl). Since salt atoms are hypothesized to be the most stable or resonant forms of matter, it is speculated to allow the salt's structure to 'tune into' or interact with the ZPE field more effectively than less ordered substances. Salts are the key to conveyance of nutrients, but now the cymatic sound vibrations are arranging their highly ordered, repeating atomic lattice into deeper coherence.

Ringing a Nepalese singing bowl before meals creates a powerful, specific frequency and complex harmonic structure that could be useful for absorption of nutrients. The sound waves (cymatics) act as the triggering vibration. A chain reaction takes place whereby the sound wave imparts a resonant frequency to the salt's crystalline structure. This resonance is hypothesized to enhance the salt's (already stable) connection or mediation with the local ZPE field.

In real life, atoms are always moving, and actual crystals contain imperfections, such as vacancies (missing atoms) or dislocations (misaligned planes of atoms), which are points of higher energy and disorder. The highly ordered, rhythmic energy of the cymatic wave could match a resonant frequency of the disordered regions. By doing so, the vibration wouldn't just add random thermal energy; it provides the precise, minimal energy needed for the misaligned atoms to snap into their correct, lowest-energy (most 'lattice-like') positions. This process is analogous to 'acoustic annealing', where controlled vibrations are used to relieve stress or improve the structure of materials. Cymatics would allow the NaCl crystal to achieve a state of higher structural coherence with fewer defects, a higher degree of structural perfection.

The mechanism hereby proposed is that the enhanced interaction with the ZPE field–facilitated by the resonant salt–creates a localized energy or coherence effect in the meal impacting nutrient absorption. This ZPE-mediated coherence is then speculated to alter the energetic state or bioavailability of the nutrients, making the food (and the ions/minerals, including salt itself) more readily absorbed by the body on a fundamental, quantum or energetic level.

The Return to Salt

We come from, and we return to, salt. In an age of a huge variety of every consumable commodity, salts, from the simple, elemental kitchen necessity, to complex healing compounds, seem to be the ultimate substance. Salts have been a foundational component, and a primal currency that, unlike the engineered novelty of synthetic chemicals under patent or the centralized oil monopolies, are scattered across the world with democratic abundance. We find their pleasure on every table and in every physiological process, a delight so ubiquitous, it is often forgotten, yet whose absence would immediately unravel the sophisticated machinery of the body and the planet.

These foundational crystalline structures of the Earth and the very regulators of our cellular function, remind us of health's elemental simplicity; their balanced presence, critical to every nerve impulse and fluid exchange, represents innate, natural efficacy. This simple, vital balance stands in contrast to a contemporary medical system whose architecture of insurance often ignores the subtle, low-cost regulatory needs of the body, prioritizing complex, patented chemical fixes.

Governments that have adopted expanded reimbursement mandates to actively promote and value holistic health and preventive care will tend to produce healthier populations, and ultimately healthier countries. By financially recognizing therapies rooted in personal responsibility and self-sustaining wellness, countries can elevate the perceived value of a life built on innate vitality and long-term well-being, rather than one defined by continuous, expensive industrial medical interventions. Instead of going against the tides of nature, we should actively support and incentivize transparent care models that pursue genuine, lasting cures and reward patient independence, letting the cyclical, self-consuming economy built on perpetual chronic management crumble away.

The soul of salts lies in their unassuming power. Whereas more spectacular and often profitable compounds—from hormones to patented molecules—are celebrated for their potency and patentability, salts operate as the silent, essential background. Their simple, crystalline truth is the very medium that holds the complex fuels and fluids of life in balance. To doubt the efficacy of these primal minerals is to miss the greater part of economic history, and the meaning of elemental chemistry. Nature's subtle perfection, salts ensure the continuation of organic performance, validating the fundamental act of survival itself. Salts are the crystalline backbone of life.

Bibliography

Abrahams, P. (2021). *Geophagia.* Wikipedia. [Online] Available at: https://en.wikipedia.org/wiki/Geophagia (Accessed: 2 February 2025).

Abrahams, P. (2021). *Mud Cookies.* Wikipedia. [Online] Available at: https://en.wikipedia.org/wiki/Mud_cookie (Accessed: 4 June 2025).

Adlam, G. . J. (2014). *Acids, Alkalis and Salts.* [eBook] Project Gutenberg.

Al-Razi (Rhazes) (925). *The Comprehensive Book on Medicine (Kitab al-Hawi fi al-Tibb).*

Ambrose, C. (2016) *Salt solutions and electrical conductivity.* July 30. Available at: https://www.youtube.com/watch?v=LqNm-bCn3nI (Accessed: 29 September 2025).

Anderson, O.E. Jr. (2015) *Refrigeration in America: A History of a New Technology and Its*

Impact. Princeton, NJ: Princeton University Press.

Arhin, A.M. and Zango, S.S. (2017). "Physicochemical and mineralogical composition studies of clays from Share and Tshonga areas, Northern Bida Basin, Nigeria: Implications for Geophagia." *Journal of African Earth Sciences*, 136, pp. 24–33.

Aristotle (1936). *Problems.* Translated by W. S. Hett. Cambridge, MA: Harvard University Press
Asare-Bediako, J. (2011) "Economic Potential of Salt Mining in Ghana Towards the Oil Find," Delve Database, 10 August. Available at: (https://www.jci.org/kiosks/access/institutional-access-troubleshooting) (Accessed: 9 May 2025).

Atis Rezistans Ghetto Biennale (2022). *Bonbon Tè Majik - Bake Sale!.* Documenta Fifteen. [Online] Available at: https://documenta-fifteen.de/en/calendar/bonbon-te-majik-bake-sale/ (Accessed: 14 January 2024).

Baldwin, J. (1956) *Giovanni's Room.* New York: Dial Press.

Barrett, S. ([no date]). *Homeopathy: The Ultimate Fake.* Available at: https://bionmr.unl.edu/courses/chem221/information/homeopathy (Accessed: 12 September 2025).

Bastin, R.J., Bowker, M.J. and Slater, B.J. (2000) "Salt Selection and Optimisation Procedures for Pharmaceutical New Chemical Entities," *Organic Process Research & Development,* 4(5), pp. 1271-1278.

Bhikha, R.A.H. and Haq, M.A. (2001). *Tibb - Traditional Roots of Medicine in Modern Routes to Health.* South Africa: Mountain of Light
The Bible. (1999) King James Version. Oxford: Oxford University Press.

Boddy, E. M. (2014). *The history of salt: With observations on its geographical distribution, geological formation, and medicinal and dietetic properties.* [eBook] Project Gutenberg.

Boericke, W. and Dewey, W.A. (1899). *The Twelve Tissue Remedies of Schüßler.* Philadelphia: Boericke & Tafel. Available at: Project Gutenberg (Accessed: 12 September 2025).

Boiron.ca. ([no date]). *Schussler salts: what is it?* Available at: https://www.boiron.ca/en/all/others/homeopathy (Accessed: 12 September 2025).

Bomar, S. (2013). *Astro-Chemico-Physiological*

and Chromatic Chart. [Online]. Available at: http://www.resistance2010.com/page/biochemical-cell-salts (Accessed: 12 September 2025).

Boye, K. (1924) *Dikter.* Stockholm: Albert Bonniers Förlag.

Brodhead, P. (2001). *The 12 Tissue Salts or Cell Salt Remedies: Fundamental homeopathic remedies.* A lecture presented by Peter Brodhead. [Online]. Available at: https://www.tibb.co.za/articles/Tissue_Salts_and_the_Qualities_of_Tibb.pdf (Accessed: 12 September 2025).

Brown, T. E., LeMay, H. E., Bursten, B. E., Murphy, C. J., Woodward, P. M. and Stoltzfus, M. E. (2023) *Chemistry: The Central Science.* 15th edn. Hoboken, NJ: Pearson.

Bundesministerium für Gesundheit (German Federal Ministry of Health). (2016). Information on Voluntary Benefits and the GKV-Modernisierungsgesetz (SHI Modernization Act) 2003. Cited in Siggelkow, M. & Leimbach, T. (2024).

Bundesgesundheitsblatt - Gesundheitsforschung - Gesundheitsschutz. [Online]. Available at: https://www.springermedizin.de/prescriptions-of-homeopathic-remedies-at-the-expense-of-the-germ/50562858 (Accessed: 14 October 2025). DE

Bundesamt für Gesundheit (BAG) / Federal Office of Public Health. (n.d.). Benefits and tariffs: Compulsory health insurance. [Online]. Available at: https://www.bag.admin.ch/bag/en/home/versicherungen/krankenversicherung.html (Accessed: 14 October 2025).

Calvert, A. F. (1919). *Salt and the salt industry.* London: Isaac Pitman & Sons. [from Calvert, A. F. (2022). *Salt and the salt industry.* eBook] Project Gutenberg.

Card, D.R. (2004). *Facial Diagnosis of Cell Salt Deficiencies.* China: Kalindi Press.

Carey, G. W. (2010). *Chemistry of Human Life.* USA: Kessinger Publishing, LLC.

Carey, G.W. (2007). *The Biochemic System of Medicine. The Theory, Pathological Action, Therapeutical Application, Materia Medica and Repertory of Schuesslers Twelve Tissue Remedies.* [Online]. Available at: http://archive.org/stream/biochemicsystemo1894care/biochemicsystemo1894care_djvu.txt (Accessed: 12 September 2025).

Carey, G.W. (2012). *The Biochemic System of Medicine. Comprising the Theory, Pathological Action, Therapeutical Application, Materia Medica and Repertory of Schuesslers Twelve Tissue Remedies.* USA: Forgotten Books.

Catedral de Sal de Zipaquira´ (2024). [Online]. Available at: https://www.catedraldesal.gov.co/ (Accessed: 12 October 2025).

Cather, W. (2006). *O Pioneers!* [eBook] Project Gutenberg.

CBS News. (2008). "Desperate Haitians Survive on Mud Cookies." [Online] Available at: https://www.cbsnews.com/news/desperate-haitians-survive-on-mud-cookies/ (Accessed: 14 January 2024).

Chishti, G.M. (1991). *The Traditional Healer's Handbook: A Classical Guide to the Medicine of Avicenna.* USA: Healing Arts Press.

Chua, L. (2017). *Cai Lan Chang Tan: Rensheng Pian (蔡瀾常談：人生篇).* Taipei: Yuan-Liou Publishing Co. Ltd.

Cleveland Clinic. (2022). *Electrolyte Imbalance.* Available at:

https://my.clevelandclinic.org/health (Accessed: 12 September 2025).

Coggins, C. L. (2012) *Pool & Spa Water Chemistry: A Comprehensive Guide.* Atlanta, GA: IPSSA, pp. 115-118.

Cooper, J. F. (2006). *The Last of the Mohicans: A Narrative of 1757.* [eBook] Project Gutenberg.
Cowan, R.S. (1983). *More Work for Mother: The Ironies of Household Technology from the Open Hearth to the Microwave.* New York: Basic Books.

Creightmore, R. (2012). *Geopathic Stress.* Land & Spirit. [Online]. Available at: http://www.landandspirit.net/html/body_geopathic_stress.html (Accessed: 12 September 2025).

Cronon, W. (1991). *Nature's Metropolis: Chicago and the Great West.* New York: W. W. Norton & Company.

Defoe, D. (2008). *A Journal of the Plague Year.* [eBook] Project Gutenberg.

Defoe, D. (2008). *Robinson Crusoe.* [eBook] Project Gutenberg.

Diaz, Dr. Jorge S. (2025). *The origin of a bizarre*

quantum phenomenon. [Video]. Available at: https://www.youtube.com/watch?v=DgrOm5nsm98 (Accessed 29 September 2025).

Dickson, T. S. (2013). *The Salt of the Earth.* [eBook] Project Gutenberg.

DiNicolantonio, J.J. (2017) The Salt Fix: Why the Experts Got It All Wrong–and How to Save Your Health by Eating More. New York: Harmony Books.

Diouf, P. S. and Diouf, A. (2020) *Traditional solar salt production in Lower Casamance (Senegal): Environmental and socio-economic benefits and disadvantages,* Scientific Research and Essays, 15(1), pp. 1-11. Available at: https://www.google.com/search?q=https://academicjournals.org/journal/SRE/article-abstract/78E000966072 (Accessed: 9 October 2025).

Diouf, I. et al. (2025) *Senegal's Iodine Puzzle: Iodine Status, Salt Iodization, and Dietary Iodine Sources,* Current Developments in Nutrition, 9(5):106008. Available at: https://www.google.com/search?q=https://doi.org/10.1016/j.cdnut.2025.106008 (Accessed: 9 October 2025).

Dirt Candy NYC. (2024). “Dinner Menu.” [Online] Available at: https://www.dirtcandynyc.com/dinner-menu (Accessed: 15 January 2024).

Dollars and Sense. (2013). “How to Make Mud Cookies.” [Online] Available at: https://www.dollarsandsense.org/how-to-make-mud-cookies/ (Accessed: 17 January 2024).

Doyle, A. C. (2008). *The Hound of the Baskervilles.* [eBook] Project Gutenberg.

Dreiser, T. (2008). *Sister Carrie.* [eBook] Project Gutenberg.

DSIR.gov.in. (2019). *ISM_H_Drugs.* Available at: https://www.dsir.gov.in/sites/default/files/2019-10/ISM_H_Drugs.pdf (Accessed: 12 September 2025).

EBSCO. ([no date]). *Biochemic tissue salt therapy.* Available at: https://www.ebsco.com/research-starters/health-and-medicine/biochemic-tissue-salt-therapy https://www.ebsco.com/research-starters/health-and-medicine/biochemic-tissue-salt-therapy (Accessed: 12 September 2025).

Eckermann, J.P. (1850) *Conversations of Goethe*

with Eckermann and Soret. Translated by S.M. Fuller. Boston: Hilliard, Gray, and Company.

Edwards, S.A. (2013). *Rudolpg Virchow, the father of cellular pathology.* [Online]. Available at: http://membercentral.aaas.org/blogs/scientia/rudolphvirchow-father-cellular-pathology (Accessed: 12 September 2025).

ElectroChem Ghana Limited (2023) Songor salt project: $88 million invested so far - Atta Akyea, Ghana Web, 21 November. Available at: https://www.ghanaweb.com/GhanaHomePage/business/Songor-salt-project-88-million-invested-so-far-ndash-Atta-Akyea-1884581 (Accessed: 9 October 2025).

Eliot, G. (2006). *Middlemarch.* [eBook] Project Gutenberg.

Ellis, S. (2011). *Biochemic Tissue Salts - Cellular Nutrition.* [Online]. Available at: http://www.frontshop.co.za/biochemic-tissue-salts-cellular-nutrition/ (Accessed: 12 September 2025).

Emerson, R. W. (2006). *Essays.* [eBook] Project Gutenberg.

Environmental Protection Agency (EPA). (1995).

"Edible clay sourcing and purification." [Online] Available at: https://www.epa.gov/sites/production/files/2016-01/documents/ap42_ch11s25.pdf (Accessed: 14 January 2024).

Environmental Protection Agency (EPA). (1995). "Fuller's Earth." [Online] Available at: https://www3.epa.gov/ttnchie1/ap42/ch11/final/c11s25.pdf (Accessed: 12 January 2024).

EP Engineered Clays. (2024). "Edible clay sourcing and purification." [Online] Available at: https://epengineeredclays.com/ (Accessed: 17 January 2024).

European Academies' Science Advisory Council (EASAC). (n.d.). European Academies' Science Advisory Council (EASAC). [Online]. Available at: https://easac.eu/ (Accessed: 14 October 2025).

European Committee for Homeopathy (ECH). (n.d.). European Committee for Homeopathy (ECH). [Online]. Available at: https://homeopathyeurope.org/ (Accessed: 14 October 2025).

European Committee for Homeopathy (ECH). (n.d.). Regulation and Reimbursement Status in

European Countries. [Online]. Available at: [https://www.inami.fgov.be/en] (Accessed: 14 October 2025).

EV Infrastructure News (n.d.) CATL's Sodium-Ion EV Battery Passes China's New Certification with 15-Minute Fast Charging Capability. Available at: https://www.evinfrastructurenews.com/ev-technology/catl-s-sodium-ion-ev-battery-passes-china-s-new-certification-with-15-minute-fast-charging-capability (Accessed: 15 October 2025).

Faraco, G. et al. (2018) "Dietary salt promotes neurovascular and cognitive dysfunction," *Nature,* 559(7713), pp. 248–252. doi: 10.1038/s41586-018-0244-7.

Farsad, V., Eckford, A.W. and Hatzinakos, D. (2013) "*Molecular communication in nanonetworks: the potential of engineered bacteria,*" IET Communications, 7(10), pp. 1003–1011. doi: 10.1049/iet-com.2012.0838.

Farthing, M. J. G. et al. (2008). "*Oral rehydration therapy: an update on the current clinical evidence,*" Clinical Gastroenterology and Hepatology, 6(11), pp. 1205-1215. (Accessed: 1 June 2025).

Food and Drug Administration (FDA). (2009). *About Dental Amalgam Fillings.* U.S. Food and Drug Administration (FDA). [Online]. Available at: http://www.fda.gov/medicaldevices/productsandmedicalprocedures/dentalproducts/dentalamalgam/ucm171094.htm (Accessed: 1 June 2025).

Food and Drug Administration (FDA). (2025). "CPG Sec 545.400 Pottery (Ceramics)." [Online] Available at: https://www.fda.gov/media/76508/download (Accessed: 1 June 2025).

Food and Drug Administration (FDA). (2024). "Environmental Contaminants in Food." [Online] Available at: https://www.fda.gov/food/chemical-contaminants-pesticides/environmental-contaminants-food (Accessed: 1 June 2025).

Food and Drug Administration (FDA). (2025). "CLAY (KAOLIN)." (HFPAPP). [Online] Available at: https://www.hfpappexternal.fda.gov/scripts/fdcc/index.cfm?set=IndirectAdditives&id=CLAYKAOLIN (Accessed: 1 June 2025).

Food Hacks (WonderHowTo). (2015). "High-end restaurants' edible 'dirt' menu." [Online] Available at:

https://food-hacks.wonderhowto.com/how-to/make-dirt (Accessed: 14 January 2024).

Food Hacks (WonderHowTo). (2015). "Make Dirt You Can Actually Eat." [Online] Available at: https://food-hacks.wonderhowto.com/how-to/make-dirt (Accessed: 17 January 2024).

Food Republic. (2011). "Dirt Is What For Dinner?" [Online] Available at: https://www.foodrepublic.com/2011/09/23/dirt (Accessed: 12 January 2024).

Fontaine, O., Gore, S.M., and Pierce, N.F. (1993). "*Rice-based oral rehydration solution for treating diarrhoea: a meta-analysis of 11 clinical trials*," The Lancet, 342(8870), pp. 493-495.

Gaskell, E. (2008). *North and South.* [eBook] Project Gutenberg.

The Fox. (2024). *Bizarre Dirt-Eating Trend Goes Viral On TikTok.* WGAR. [Online] Available at: https://www.bing.com/search?q=Anon.+(2021).+Bizarre+Dirt-Eating+Trend+Goes+Viral+On+TikTok.+WGAR&cvid=5c27d53e579c42eeb84346c4983ce8fb&gs_lcrp=EgRlZGdlKgYIABBFGDkyBggAEEUYOTIHCAEQ6wcYQNIBBzg3N2owajSoAgiwAgE&FORM=ANAB01&PC=U531 (Accessed: 14 August 2025).

GEPA Exporters Portal (2024) "US$1billion annual salt revenue by 2028," GEPA News, 13 September. Available at: https://owl.purdue.edu/owl/general_writing/grammar/using_articles.html (Accessed: 9 October 2025).

Giornale Italiano di Nefrologia. (2018). *The historical relevance of urine and the future implications.* Available at: https://giornaleitalianodinefrologia.it/en/2018/01/the-historical-relevance-of-urine (Accessed: 12 September 2025).

Gissing, G. (2004). *The Private Papers of Henry Ryecroft.* [eBook] Project Gutenberg.

GKV-Spitzenverband (National Association of Statutory Health Insurance Funds). (n.d.). Statutory health insurance. [Online]. Available at: https://www.gkv-spitzenverband.de/english/statutory_health (Accessed: 14 October 2025).

Glahn, R. P., Young, S. L. and Tako, E. (2014) "Bioavailability of iron in geophagic earths and clay minerals, and their effect on dietary iron absorption using an in vitro digestion/Caco-2 cell model," PLoS ONE, 9(5), p. e97621. Available at: https://doi.org/10.1371/journal.pone.0097621 (Accessed: 4 November 2025).

Grey, A. (2004). *What You Need To Know About The Energy Field.* Metaphysical Solutions. [Online]. Available at: http://www.metaphysicalsolutions.com/energyfield.htm (Accessed: 12 September 2025).

Guerrant, R. L. et al. (2003). "Practice guidelines for the management of infectious diarrhea," Clinical Infectious Diseases, 36(8), pp. 997–1004. doi:10.1086/368173.

Haggard, H. R. (2007). *King Solomon's Mines.* [eBook] Project Gutenberg.

Hanks True BBQ (no date) *Lär dig saltlag - Hank's True BBQ™.* Available at: https://hankstruebbq.com/se/lar-dig-saltlag/ (Accessed: 16 September 2025).

Hanks True BBQ (no date) *Lär dig saltlag - Hank's True BBQ™.* Available at: https://hankstruebbq.com/se/lar-dig-saltlag/ (Accessed: 16 September 2025).

Hanson, N. (2017). "Evaluation of the role of povidone-iodine in the prevention of surgical site infections," *International Journal of Environmental Research and Public Health,*

14(6), p. 642. doi:10.3390/ijerph14060642.

Healthline. (2023). "Bentonite Clay." [Online] Available at: https://www.healthline.com/health (Accessed: 12 January 2024).

Health Tips & Tricks. (2023). *2 Vitamins You Should Take At Night To Grow Muscle (Scientifically Proven).* [Video]. 22 November 2023. Available at: https://www.youtube.com/watch?v=c-cXN8Q2rMw (Accessed: 11 January 2024).

Healthline. (2024). *Tissue Salts.* Available at: https://www.healthline.com/health (Accessed: 12 September 2025).

Henderson, A. (2009). *Tissue Salts - Health 24.* [Online]. Available at: http://www.health24.com/Natural/Therapies/Tissue-salts-20120721 (Accessed: 12 September 2025).

Henry's Health Habits (2025) *No. 1 Weight Loss Expert Reveals: SALT Switches On FAT BURNING After 40 (Dr. Ben Bikman).* Available at: https://www.youtube.com/watch?v=c-cXN8Q2rMw (Accessed: 16 September 2025).

Herodotus. *The Histories.* Translated by Robin

Waterfield. Oxford University Press, 1998.

Homer, 2000. *The Odyssey,* translated by R. Fagles, London: Penguin Classics.

The Homoeo Ocean (2024) *The Biochemic system of medicine & Twelve tissue salts by Dr. Schussler (Part - 1).* Online video. Available at: https://www.youtube.com/watch?v=U-NU3vhghPU (Accessed: 3 October 2025).

Hopkins, C. 2015. "Tissue Salts." [Online] Available at: https://www.scribd.com/document/313252987/Tissue-Salts-Combinations (Accessed: 20 March 2025).

The Human Condition (2022) *Micro Dosing Supplements,* [Video], Available at: http://www.youtube.com/watch?v=EG01R0IxhUU (Accessed: 17 May 2025).

Huan Kuan (桓寬) (Han Dynasty) *Yán Tiě Lùn* (鹽鐵論). In: Sìbù Cóngkān Chūbiān (四部叢刊初編) [First Series of the Sibu Congkan]. Shanghai: Commercial Press.

桓寬 (Huan, K.) (1985). *The Discourses on Salt and Iron: A Debate on State Control of Commerce and Industry in Ancient China.*

Translated by E. M. Gale. Taipei: Ch"eng Wen Publishing Company.

Hussain, M. (2013) "Senegal looks to the salt of the Earth - in pictures," *The Guardian,* 29 March. Available at: https://www.theguardian.com/global-development/gallery/2013/mar/29/senegal-salt-earth-in-pictures (Accessed: 4 February 2025).

Ibn Sina (Avicenna) (1025). *The Canon of Medicine (Qanun fi al-Tibb).*

IGI Global. (2025). *The Biochemic System of Medicine.* Available at: https://www.igi-global.com/viewtitle.aspx?TitleId=207987 (Accessed: 12 September 2025).

IJAR. (2018). *A Review on Therapeutic Uses of Cow Urine.* Available at: https://www.journalijar.com/uploads/2018/08/114_IJAR-24301.pdf (Accessed: 12 September 2025).

IMANI Africa (2024) "Prospects of Ghana's Salt Industry: A Pathway to adding an extra $1 Billion in foreign exchange earnings," IMANI Policy Brief, September. Available at: https://imaniafrica.org/2024/09/prospects-of-ghanas-salt-industry-a-pathway-to-adding-an-extra-1-billion-in-foreign-exchange-earnings/

(Accessed: 9 August 2025).

INAMI/RIZIV (National Institute for Health and Disability Insurance, Belgium). (2016). "Healthcare refunds: Which types of healthcare are reimbursed by the Belgian compulsory health insurance system?" [Online] Available at: https://www.inami.fgov.be/en (Accessed: 14 October 2025).

Invest for Jobs (2023) "First Salt Industry Business Forum in Senegal," Invest for Jobs News, 22 May. Available at: https://invest-for-jobs.com/en/news/first-salt-industry-business-forum-in-senegal (Accessed: 9 June 2025).

Iran Bentonite Co. (2024) 'Bentonite Dangers', Iran Bentonite Co. Available at: https://iranbentoniteco.com/bentonite-dangers/ (Accessed: 4 March 2025).

IRDAI (Insurance Regulatory and Development Authority of India). (2024).

Iodine Global Network (IGN) (no date) *Senegal.* Available at: https://ign.org/latest/newsletter/senegal-struggles-to-control-iodine (Accessed: 9 August 2025).

James, R.W., Waller, I. and Hartree, D.R.

(1928) "An investigation into the existence of zero-point energy in the Rock-Salt Lattice by an X-Ray Diffraction Method," Proceedings of the Royal Society of London. Series A, Containing Papers of a Mathematical and Physical Character, 118(779), pp. 334–350.

Jogani, K. (no date) *Speculative Technologies.* [Presentation] Scribd. Available at: https://www.scribd.com/presentation/719364844/Speculative-Technologies-Kishan-Jogani (Accessed: 29 September 2025).

Kevorkian, J. (1960). "Rudolph Virchow," JAMA, Vol 172, No. 15. [Online]. Available at: http://jama.jamanetwork.com/article.aspx?articleid=328355 (Accessed: 12 September 2025).

Kansas Farm Food Connection. (2024). "How to Make Edible Soil." [Online] Available at: https://kansasfarmfoodconnection.org/spotlights/go-ahead-%E2%80%94-eat-some-dirt (Accessed: 14 January 2025).

Kopp, J. (2012). *Various Effects of Underground Water as Illustrated.* [Online]. Available at: http://www.landandspirit.net/html/body_geopathic_stress.html (Accessed: 12 September 2024).

Kovach, S. (2007). *The Hidden Dangers of Cell*

Phone Radiation. Life Extension Magazine LE Magazine. [Online]. Available at: http://www.lef.org/magazine/mag2007/aug2007_report_cellphone_radiation_01.htm (Accessed: 1 September 2024).

Kurlansky, M. (2002). *Salt: A world history.* New York: Walker and Co.

Lamoreaux, S.K. (1997). *Demonstration of the Casimir Force in the 0.6 to 6μm Range.* Physical Review Letters, 78(1), pp. 5–8. Available at: https://doi.org/10.1103/PhysRevLett.78.5 (Accessed 30 September 2025).

Le Bon La Butte. (2024). "Menus." [Online] Available at: https://www.lebonlabutte.fr/en/menus/ (Accessed: 17 January 2025).

Laozi (1963). *Tao Te Ching.* Translated by D. C. Lau. London: Penguin Books.

Li, W. (2024) Seniors, Stop Taking K2+D3 COMBO Pills - They're Destroying Your Kidneys [Video]. YouTube. Available at: https://www.youtube.com/watch?v=BB65or3TVBU (Accessed: 20 October 2025).

Livy. *The History of Rome, Books 1-5.*

Translated by Valerie M. Warrior. Hackett Publishing Company, 2006.

Majno, G. (1991). *The Healing Hand: Man and Wound in the Ancient World.* Cambridge, MA: Harvard University Press.

Maugham, W. S. (2007). *Of Human Bondage.* [eBook] Project Gutenberg.

Medical News Today. (2023). "The Truth About Eating Dirt for Health." [Online] Available at: https://www.medicalnewstoday.com/articles/eating-dirt (Accessed: 17 January 2024).

Meehan, J. (2016). *Cymatics Video: Salt on Steel Plate played with bow by Jodina Meehan Video.* YouTube. Available at: https://www.youtube.com/watch?v=YkRC2YmHFn4 (Accessed 3 September 2024).

Melville, H. (2004). *Moby Dick; Or, The Whale.* [eBook] Project Gutenberg.

Mexicolore. (2009). *Clean Aztecs, Dirty Spaniards.* Available at: https://www.mexicolore.co.uk/aztecs/home/clean-aztecs-dirty-spaniards (Accessed: 1 September 2024).

Milton, J. (2007). *Paradise Lost.* [eBook] Project Gutenberg.

MJPMS.in. (2014). *Indian cow urine distillation and therapeutic uses.* Available at: https://www.mjpms.in/articles/indian-cow-urine-distillation-and-therapeutic-uses.pdf (Accessed: 1 December 2024).

Moberg, V. (1956) *Detta är min första bok.* Stockholm: Bonniers.

Moe, S., Baker, J., Lydon, T. C., et al. (2022). "Drinking water chlorination has minor effects on the intestinal flora and resistomes of Bangladeshi children." *Nature Microbiology,* 7(6), 842–853.

Moleschott, J. (1852) *Kreislauf des Lebens (The Cycle of Life).* Mainz: Victor von Zabern.

Mulliken, R.S. (1925) *The Isotope Effect in Band Spectra, II: The Spectrum of Boron Monoxide,* Physical Review, 25(3), pp. 259–294. doi: 10.1103/PhysRev.25.259.

National Center for Biotechnology Information (NCBI). (2010). *Cow urine distillate as bioenhancer.* Available at: https://pubmed.ncbi.nlm.nih.gov/21731367/ (Accessed: 12 September 2025).

Nietzsche, F. (2007). Thus Spake Zarathustra. [eBook] Project Gutenberg.

NTRS. (1982). "Prolonged triboluminescence in clays and other minerals." [Online] Available at: https://ntrs.nasa.gov/citations/19820049748 (Accessed: 17 January 2024).

The Nutritional Institute. (2014). "A focus on tissue salts." Available at: https://thenutritionalinstitute.com/resources/articles/139-a-focus-on-tissue-salts (Accessed: 12 September 2025).

Opoku, K.A. (1997). *Hearing and Keeping: Akan Proverbs.* Accra: Adwinsa Publications.

Paris Zig Zag. (2024). "Top des bouis-bouis parisiens où l'on se régale." [Online] Available at: https://www.pariszigzag.fr/bar-restaurant/restaurant/top-des-bouis-bouis-parisiens-ou-lon-se-regale (Accessed: 17 January 2025).

Paulekuhn, G. S., Löbmann, K. A. and Rades, T. (2014) "Pharmaceutical salts of small molecule drugs: opportunities and challenges," *European Pharmaceutical Review,* 19(6), pp. 10–18. Available at:

https://www.europeanpharmaceuticalreview.com/article/27753/pharmaceutical-salts-small-molecule-drugs/ (Accessed: 10 October 2025).

Permies.com. (2013). Removing Salt from Urine. Available at: https://permies.com/t/22926/Removing-Salt-Urine (Accessed: 12 January 2025).

Pester, A. (2014). "A restaurant in Japan is serving a $110 tasting menu featuring dirt," *Smithsonian Magazine.* Available at: https://www.smithsonianmag.com/smart-news/a-restaurant-in-japan-is-serving-a-110-tasting-menu-featuring-dirt (Accessed: 14 September 2025).

Petit Bouillon Pharamond. (2024). "Petit Bouillon Pharamond Menu." [Online] Available at: https://www.petitbouillonpharamond.com/the-restaurant-menu-1 (Accessed: 17 January 2025).

Pliny the Elder. *Natural History: A Selection.* Translated by John F. Healy. Penguin Books, 1991.

Poe, E. A. (2006). *The Narrative of Arthur Gordon Pym of Nantucket.* [eBook] Project Gutenberg.

Project Learning Tree. (2024). "Edible Recipes to Teach Students about Soil." [Online] Available at: https://www.plt.org/educator-tips/edible-recipes-teach-students-soil/ (Accessed: 19 January 2025).

Protoleaf, Inc. (no date). *Products.* Available at: https://www.protoleaf.co.jp/products/ (Accessed: 14 September 2025).

Purdue University. (2016). "How to Make Edible Soil." [Online] Available at: https://ag.purdue.edu/department/asec/_docs/natural_resources/ediblesoilprofilewolf.pdf (Accessed: 14 January 2024).

ReAgent Chemicals (no date). "Types & Uses of Salts in Chemistry." *The Science Blog.* Available at: https://www.reagent.co.uk/blog/types-of-salts-in-chemistry-and-their-uses/ (Accessed: 12 September 2025).

ResearchGate. (2015). "Geophagy as a risk factor for helminth infections in pregnant women in Lusaka, Zambia." [Online] Available at: https://www.researchgate.net/publication/272470947_Geophagy_As_A_Risk_Factor_For_Helminth_Infections_In_Pregnant_Women_In_Lusaka_Zambia (Accessed: 17 January 2024).

ResearchGate. (2010). "Parasite transmission risk from geophagic and foraging behaviour in chacma baboons." [Online] Available at:

https://www.researchgate.net/publication/227176950_Parasite_Transmission_Risk_From_Geophagic_and_Foraging_Behavior_in_Chacma_Baboons (Accessed: 17 January 2024).

ResearchGate. (2007). "The potential impact of soil ingestion on human mineral nutrition." [Online] Available at: https://www.researchgate.net/publication/8349208_The_potential_impact_of_soil_ingestion_on_human_mineral (Accessed: 17 January 2024).

ResearchGate. (2021). "Clay Minerals Affect the Solubility of Zn and Other Bivalent Cations in the Digestive Tract of Ruminants In Vitro." [Online] Available at: https://www.researchgate.net/publication/350209617_Clay_Minerals_Affect_the_Solubility_of_Zn_and_Other_Bivalent_Cations_in_the_Digestive_Tract_of_Ruminants_In_Vitro (Accessed: 14 January 2024).

ResearchGate. (2007). "Wilhelm Heinrich Schussler: A therapist of his era." Available at: https://www.researchgate.net/publication/290793474_Wilhelm_Heinrich_Schussler_A_therapist_of_his_era (Accessed: 12 September 2025).

ResearchGate. (2025). "Urine therapy in Ayurveda: A time-honoured tradition for health and healing." Available at: https://www.researchgate.net/publication/389430911_Urine_therapy_in_Ayurveda_A_time-_honoured_tradition_for_health (Accessed: 12 September 2025).

Roberts, M. (2005). *Tissue Salts for Healthy Living.* SA: Spearhead -An imprint of New Africa Books (Pty) Ltd.

Roberts, M. (2011) *Tissue Salts for Healthy Living.* Cape Town: Penguin Random House South Africa.

Root-Bernstein, M. and Root-Bernstein, R. (2000). *Honey, Mud, and Slime: The Wonders of Geophagy and Other Earth-Eating Habits.* Houghton Mifflin Harcourt.

Rosen, G. (1993). *A History of Public Health.* Baltimore: The Johns Hopkins University Press.
Rubens, P.P. (1638). *Picture of the statue of Hippocrates.* [Online]. Available at: http://upload.wikimedia.org/wikipedia/commons/3/32/Hippocrates_rubens.jpg (Accessed: 11 February 2025).

Sabat, D., Dziembala, A. & Panasiewicz, M. (2002). "Rudolf Virchow and presentation of his scientific achievement in Polish medical magazines in the 19th century and the beginning of 20th century," (Pol J Pathol. 2002;53(3):163-8). [Online]. Available at: http://www.ncbi.nlm.nih.gov/pubmed/12476619 (Accessed: 12 May 2025).

Saunders. (2007). *Dorland's Illustrated Medical Dictionary.* USA: Elsevier.

Savica, V., Calò, L. A., Santoro, D., Monardo, P., Mallamace, A. & Bellinghieri, G. (2011) "Urine therapy through the centuries." *Journal of Nephrology,* 24 (Suppl 17), pp. S123–S125.

Schivelbusch, W. (1992). *Tastes of Paradise: A Social History of Spices, Stimulants, and Intoxicants.* New York: Vintage Books.

Schoenfeld, E.F. (2009). Second Chance. *Regain Your Health With Tissue Salts.* SA: Graysonian Press.

Schoenfeld, E.F. (2013). *Psychosomatic Connections.* Level 3/No 1-6.Manual. SA: Academy of Tissue Salts & Facial Analysis (ATIFA).

Schoenfeld, E.F. (2013). *Tissue Salts & Facial analysis.* Level 1-12 Manual. SA: Academy of Tissue Salts & Facial Analysis (ATIFA).

Schuessler, W.H. (2012). *An Abbreviated Therapy: The Biochemical Treatment of Disease.* USA: Ulan Press.

Schuessler, Dr. Med., (1898). *An abridged therapy : manual for the biochemical treatment of disease.* Boericke & Tafel, 1898. https://iiif.wellcomecollection.org/pdf/b21152925

SchuesslerTissueSalts.uk. (n.d.). *What are Schuessler Tissue Salts?* Available at: https://schuesslertissuesalts.uk/pages/what-are-schuessler-tissue-salts (Accessed: 12 March 2025).

Seim, G. L., Tako, E., Ahn, C., Glahn, R. P. and Young, S. L. (2016) 'A Novel in Vivo Model for Assessing the Impact of Geophagic Earth on Iron Status', Nutrients, 8(6), p. 362. Available at: https://doi.org/10.3390/nu8060362 (Accessed: 4 January 2025).

SemperTrue (2023). *How Salt Mines Created Underground Cities?* Available at: https://www.youtube.com/watch?v=TLJO5LmJzWo (Accessed: 13 July 2025).

SemperTrue (2023). *Why the Biggest Battery Company is Betting Against Lithium.* [Online Video]. Available at:

https://www.youtube.com/watch?v=TLJO5LmJzWo (Accessed: 14 September 2025).

Servan-Schreiber, D. (2008) *Anticancer: A New Way of Life.* New York, NY: Viking Press.

Shinondo, C.J. and Mwikuma, G. (2010). "Geophagy as a risk factor for helminth infections in pregnant women in Lusaka, Zambia," *Medical Journal of Zambia.*

SGS. (2022). "Food Contact Material Regulations - USA." [Online] Available at: https://www.sgs.com/en/services/food-contact-material-regulations-usa (Accessed: 14 January 2024).

Shinozaki, Y. and Ohashi, K. (2017). "Normal Saline (0.9% NaCl) and Lactated Ringer's Solution," in *Perioperative Fluid Management.* Tokyo: Springer, pp. 27–34.

Siciliana, N. (2013). "Schüssler Tissue Salts." [Online]. Available at: http://similia.wordpress.com/schussler-tissue-salts/ (Accessed: 12 September 2025).

Sima, Q. (1993). *Records of the Grand Historian.* Translated by B. Watson. New York: Columbia University Press.

Sneader, W. (2005). *Drug Discovery: A History.* Chichester, UK: John Wiley & Sons.

Sotiriou, E. et al. (2025) "Comparing the efficacy of chlorhexidine and povidone-iodine in preventing surgical site infections: A systematic review and meta-analysis," Journal of Evidence-Based Medicine, 18(2), pp. 100-115. doi:10.1111/jebm.12658.

Stahl, P.H. and Wermuth, C.G. (2008). *Handbook of Pharmaceutical Salts: Properties, Selection, and Use.* Weinheim, Germany: Wiley-VCH.

Strasser, S., 1982. *Never Done: A History of American Housework.* New York: Pantheon Books.

Strazzullo, P. et al. (2009) "Salt sensitivity of blood pressure: a systematic review and meta-analysis of controlled trials," *BMJ,* 339, b5220. doi: 10.1136/bmj.b5220.

Svensson, L. (2018). "Moberg's Role as Social Critic and the Stinging Salt." Available at: https://www.literaryreview.org/moberg-salt.html (Accessed: 15 October 2025).

Swenson, J. (2004). *Chakra Force Tissue*

Salt Therapy. [Online]. Available at: http://www.chakraforce.com/Salt.html (Accessed: 12 September 2025). Tibb Institute. (2012). *Tibb Training Manual. Lifestyle Advisor Training Course.* Johannesburg: Ibn Sina Institute of Tibb.

Tchobanoglous, G., Burton, F. L. and Stensel, H. D. (2014) *Wastewater Engineering: Treatment and Resource Recovery, 5th Edition.* New York: McGraw-Hill Education. pp. 789-795.

TheFork. (2024). "Le Boui Boui Paris restaurant." [Online] Available at: https://www.thefork.com/restaurant/le-boui-boui-r10033 (Accessed: 17 January 2024).

TheFork. (2024). "Le Boui Boui Paris restaurant." [Online] Available at: https://www.thefork.com/restaurant/le-boui-boui-r10033/menu (Accessed: 17 January 2024).

ThoughtCo. (2019). "Why Eating Dirt Is Still A Thing." [Online] Available at: https://www.thoughtco.com/geophagy (Accessed: 17 January 2024).

TikTokers Eating Dirt? (2023). VegNews. [Online] Available at: https://vegnews.com/tiktokers-dirt (Accessed: 17 January 2024).

Titford, M. (2009). *Rudolf Virchow: Cellular Pathologist,* Lab Med, 41(5), pp. 311-314. [Online]. Available at: http://labmed.ascpjournals.org/content/41/5/311.full (Accessed: 12 September 2025).

Torabi, H. et al. (2021) "Independent and interactive associations of dietary nitrate and salt intake with blood pressure and cognitive function: a cross-sectional analysis in the InCHIANTI study," *International Journal of Food Sciences and Nutrition,* 73(4), pp. 439–448. doi: 10.1080/09637486.2021.1993157.

Trumbull, H. C. (2014). *The Covenant of Salt: As based on the significance and symbolism of salt in primitive thought.* [eBook] Project Gutenberg.

Tsavo Trust (2025) "Geophagy: why do animals eat soil?" Available at: https://tsavotrust.org/geophagy-why-do-animals-eat-soil/ (Accessed: 4 March 2025).

Türkiye Today (2025) "Correct the Map: Africa Union pushes for maps that reflect its true size, Türkiye Today." Available at: https://www.turkiyetoday.com/world/correct-the-map-africa-union-pushes-for-maps-that-reflect-its-true-size-3206253 (Accessed: 10 October 2025).

Twain, M. (1999). *Life on the Mississippi.* [eBook] Project Gutenberg.

Twain, M. (2008). *A Connecticut Yankee in King Arthur's Court.* [eBook] Project Gutenberg.

U.S. Food and Drug Administration (FDA) (2015) "Questions and Answers: FDA requests label changes and single-use packaging for some over-the-counter topical antiseptic products to decrease risk of infection." Available at: https://www.fda.gov/drugs (Accessed: 13 September 2025).

U.S. Food and Drug Administration (FDA). (2025). "Homeopathic Products." Available at: https://www.fda.gov/drugs (Accessed: 12 September 2025).

U.S. Government (1977) "United States Geological Survey: Water Dowsing." General Interest Publication. Washington, D.C.: U.S. Government Printing Office.

USGS. (2019). "Ion Exchange in Clays and Other Minerals." [Online] Available at: https://pubs.usgs.gov/publication/70211664 (Accessed: 17 January 2024).

US National Library of Medicine. (2015). "Heavy Metals and their Toxicity Mechanisms." [Online] Available at: https://pmc.ncbi.nlm.nih.gov/articles/PMC4427717/ (Accessed: 14 January 2024).

U.S. Food and Drug Administration (FDA) (2015) "Questions and Answers: FDA requests label changes and single-use packaging for some over-the-counter topical antiseptic products to decrease risk of infection." Available at: https://www.fda.gov/drugs (Accessed: 14 June 2025).

USNRL (2014) *Creating Fuel from Seawater.* 7 April. Available at: https://www.youtube.com/watch?v=Iavz7AnKI8I (Accessed: 29 September 2025).

Valentine, J. (2013). *Soft Drinks - America.* [Online]. Available at: http://www.globalhealingcenter.com/american-trends/soft-drinksamerica (Accessed: 12 September 2025).

Varner, J. (2018) "Mythology Monday: The Power of Salt." Available at: https://www.jeremyvarner.com/blog/mythology-monday-the-power-of-salt (Accessed: 25 September 2024).

Verne, J. (2003). *The Mysterious Island.* [eBook] Project Gutenberg.

Verne, J. (2003). *Twenty Thousand Leagues Under the Sea.* [eBook] Project Gutenberg.

Wani, K., Saini, A. and Salunke, S. (2024) "Modal Frequency Prediction of Chladni Patterns Using Machine Learning," Journal of Modern Physics, 15(05), https://www.scirp.org/journal/paperinformation?paperid=134852 (Accessed: 16 September 2025).

Wayne, A. & Newell, L. (2000). *The proven dangers of microwave ovens.* The Christian Law Institute & Fellowship Assembly. [Online]. Available at: http://www.globalhealingcenter.com/health (Accessed: 12 September 2025).

WebMD. (2023). "What Is Dirt Eating?" [Online] Available at: https://www.webmd.com/mental-health (Accessed: 17 January 2024).

Webster, L., C. C. Reines, J. L. Nie, D. R. Lancaster, R. F. Clark, S. S. Shuford, and M. W. Stanard. (2016) "Cohort Study of the Impact of High-Dose Opioid Analgesics on Overdose Mortality," *Pain Medicine,* 17(1), pp. 85-94.

doi:10.1111/pme.12836.

Wells, H. G. (2006). *The First Men in the Moon.* [eBook] Project Gutenberg.
Xiao, F., Zhang, X., & Zhai, H. (2012). "New halogenated disinfection byproducts in swimming pool water and their permeability across skin." *Environmental Science & Technology,* 46(12), 6521-6530.

Wikipedia (no date). "Rudolf Virchow." [Online]. Available at: http://en.wikipedia.org/wiki/Rudolf_Virchow (Accessed: 12 September 2025).

Wikipedia (no date). *Salt.* Available at: https://en.wikipedia.org/wiki/Salt (Accessed: 12 September 2025).

Wikipedia. (2025). *Evidence and efficacy of homeopathy.* Available at: https://en.wikipedia.org/wiki/Evidence_and_efficacy_of_homeopathy (Accessed: 2 October 2025).

Wikipedia. (2024). "Nostalgie de la boue." [Online] Available at: https://en.wikipedia.org/wiki/Nostalgie_de_la_boue (Accessed: 17 January 2024).

Wolf, M. (1998). *The Salt-Cellars.* [eBook]

Project Gutenberg. Available at: https://www.gutenberg.org/files/1400/1400-h/1400-h.htm (Accessed: 17 January 2024).

World Health Organization (2006) *Oral rehydration salts: production of the new ORS.* Available at: https://iris.who.int/bitstream/handle/10665/69227/WHO_FCH_CAH_06.1.pdf?sequence=1 (Accessed: 13 September 2025).

World Health Organization (WHO) and United Nations Children's Fund (UNICEF) (2006). *Oral rehydration salts: production of the new ORS.* Available at: https://apps.who.int/iris/handle/10665/69225 (Accessed: 14 September 2025).

YouTube. (2016). Chefs serving edible soil. [Online] Available at: https://www.youtube.com/watch?v=mjbgxH8rSQM (Accessed: 4 January 2024).

INDEX

THE END

www.ingramcontent.com/pod-product-compliance
Lightning Source LLC
LaVergne TN
LVHW041057080826
845145LV00007B/1608

* 9 7 8 1 9 4 7 3 8 6 0 3 7 *